THE LIVING BODY

KARL SABBAGH

with Christiaan Barnard

THE LIVING BODY

Macdonald

in association with Channel Four Television Company Limited

A Macdonald BOOK

Copyright ©1984 Multimedia Publications (UK) Ltd

First published in Great Britain in 1984 by
Macdonald & Co (Publishers) Ltd
London & Sydney

A member of BPCC plc

British Library Cataloguing in Publication Data

Sabbagh, Karl
   The living body
   1. Human physiology
   I. Title
   612   QP34.5

   ISBN 0-356-10506-7

This book was devised and produced by
Multimedia Publications (UK) Ltd

Editors Jo Cheesewright, Anne Cope
Production Judy Rasmussen
Design Behram Kapadia
Picture research Diana Korchien, Catherine Blackie,
   Paul Snelgrove, Helena Beaufoy
Indexing Margaret Rawnsley

Origination by The Clifton Studio Ltd, London
Typesetting by Keene Graphics Ltd, London
Printed and bound in Italy by A. Mondadori, Verona

Macdonald & Co (Publishers) Ltd
Maxwell House
74 Worship Street
London EC2A 2EN

Most of us only become interested in our bodies when something goes wrong. We treat our bodies like cars: as long as they give us so many miles to the gallon, we care very little about what goes on under the bonnet. Which is a pity, because the living body is endlessly fascinating, and becoming increasingly so as we invent more and more sophisticated imaging techniques. The incredible vistas of the human body revealed by scanning and transmission electron microscopy, thermography, ultrasound and endoscopy were largely the impetus for The Living Body as a television series.

Obviously it has not been possible, even in 26 television programmes and now in this book, to convey the sum total of present knowledge about the workings of the human body. But we have explored many of its extraordinary beauties and intricacies, and some of its partly-solved mysteries. One of the mysteries I find most challenging is this: how, exactly, do our bodies recognize foreign tissues and invaders? We know some of the answers, but not all of them. When we do, the problem of rejection of transplanted or grafted tissues and organs will be overcome.

To give you some idea of the complexity of the problem, every cell in our body is identified not by one password but by three, and each of these passwords has something like 40 variants. Any cell without the correct password combination is immediately attacked and killed by special cells called lymphocytes. Yet another group of proteins on the surface of many of our body cells plays a part in deciding how strong our response to foreign cells should be. All of these recognition systems have evolved for the express purpose of keeping one jump ahead of disease organisms, but they make the job of a transplant surgeon very difficult. But there is another twist to the tale. The majority of women conceive and carry a baby – half an outsider, biologically speaking – to term without complications. So there is a very fine balance between mechanisms in the mother's blood that automatically repel invaders and more recently evolved mechanisms that suppress them during pregnancy. It is in 'ordinary' events such as conception and pregnancy that the miracle of the living body lies.

I have emerged from the two years that it has taken to make The Living Body television series with a great respect for the stamina of television producers! I am referring of course to Karl Sabbagh, who has patiently and inventively recreated the workings of the human body both for the television screen and for this book. Acting as medical adviser to the project has given me great satisfaction, because I have always taken the job of explaining the body to patients and others very seriously. I hope this book gives you a similar satisfaction.

CHRISTIAAN BARNARD

Contents

Acknowledgements

As with any book linked with a television series, the work done for the programmes fed the process of creating the book.

Goldcrest Multimedia and John Gau made the initial and most important contribution, by asking me to write and produce the Channel 4 television series, and by encouraging the approach I adopted to the human body in devising the series. I have used the same approach in this book – revealing the body through the tasks it performs, rather than through the organs and systems it contains.

The second important stimulus to the book and the television series has been the participation of Professor Christiaan Barnard who has been involved with the project since its earliest days. Since his surgical triumphs in the field of heart transplantation Christiaan Barnard has played an increasing role in informing the public about medicine and human biology, and this book reflects the importance of that task.

Much of what appears in the television series and in the book has benefited from the advice of two of the people who have made a major contribution to the television series, Cynthia Clarke, AIMBI, medical artist, and David Barlow, biological film-maker. Through their work for the series and their encouraging conversations during the two year's preparation of it, they have helped me clarify my thoughts and understand the human body better. Other members of the production team – including Thelma Rumsey, Jenny Toynbee-Holmes, Martin Weitz, Stuart Urban, Peter Mander and Cathy Devine – have also provided stimulating discussions about subject matter.

The ideas in the book, and the way they are expressed, are mine. Or rather they are other peoples', selected by me as being the most helpful in understanding how various tasks are carried out by the body.

I have included under *Sources* on page 219 a list of the main books I have found useful. Where I quote a fact or a figure, it will usually be found in at least one of those books although I have had to face the fact that there are occasionally disagreements even among physiologists. The length of the digestive tract or the number of brain cells, for example, tends to vary erratically from authority to authority. Since I have not the means to measure them myself I have sometimes had to make an arbitrary choice about which figures to give.

I have decided to risk occasional oversimplification in the hope that the book will clarify aspects of the body for people who would never normally read a 'science' book. This means that, for those who know about a topic, there may be an annoying shortage of specifics – 'another hormone', instead of naming it, 'a part of the brain', instead of saying 'the amygdaloid nucleus', and so on. In trying to simplify, I have inevitably had to omit, although I have tried to ensure that no major aspect of the workings of the body has been left out.

For giving me a great deal of advice after reading the television scripts and the manuscript of the book I would like to thank Professor Geoffrey Burnstock, F.R.S. and Dr Margaret Sumner, both of the Department of Anatomy at University College, London. They have been invaluable in their encouragement and guidance. As we all do with advice, I have tended to accept the bits I agree with and churlishly ignored the rest, so there is no sense in which they can be blamed for any errors or omissions in the book.

The others I would like to thank are Dr Norman Saunders, Reader in Physiology at University College, London, who also read both the television script and the manuscript of the book, and Professor Noel Dilly of the Department of Anatomy at St George's Hospital Medical School, London, who read the final galleys. Like any good scientist Dr Saunders has a deep suspicion of teleological theorizing about evolution and would like, I'm sure, to remove any hint of such an approach from the book. I, equally strongly, feel that suggesting what a particular organ or system was 'intended' to achieve is a useful figure of speech rather than a heretical restatement of evolution, and I still occasionally resort to it.

Finally, I should like to thank several colleagues at Multimedia for their contributions to the book. Jo Cheesewright, with the help of Kathy Rubinstein and Sydney Francis, brought order and cohesion out of an erratic flow of manuscript chapters, and Anne Cope masterminded the complex final stages of putting together text, illustrations and captions to produce a seamless whole.

KARL SABBAGH

1 · Design for Living

When biologists look at design and function, they call it anatomy and physiology, but this is not a biology textbook. Instead of exploring the body organ by organ or system by system, the chapters in this book often take as their starting point some ordinary human task, such as moving, eating, growing, choosing or reproducing. Many of these tasks involve more than one organ or system, and some of them depend on an intimate collaboration between *all* body systems.

For most of these processes, it is difficult to think of ways in which the body could have been improved. The eye, for example, works at the limits of available light and can discriminate very fine detail. Heart muscle tissue has abilities at a microscopic level which make it ideally suited to pump blood without our conscious control for the 3000 million beats or so in the average human lifetime. The gut, the ear, the skin – each of these is suited to a particular task that helps us to survive.

But, for many of us, the more fortunate members of the human race, survival is no longer a problem. We are not threatened by predators, we do not have to hunt or forage for our food, we are not put at risk by harsh climate or crippling diseases. And yet the human body contains mechanisms that were needed by our ancestors to deal with all of these threats. It is a survival machine, equipped with a wide range of devices and systems which can monitor the world on our behalf, detect danger and repair damage. The more we look at the detailed workings of the living body, the more we see how it has evolved to perform a wide range of useful tasks that many machines, even the most modern cannot match.

One warning should be given at this point. It is easy when describing the design of the body to slip into the habit of talking about the 'purpose' of various changes. But as we will see, evolution by natural selection cannot have a 'purpose'. Many changes take place and a few survive, but those few were not 'intended' to survive when they occurred.

The evolutionary process that has resulted in the human body is millions of times slower than the process of designing the best machine but, like the hare and the tortoise, the end result is much better.

The human body, or parts of it, has undeniably evolved: its 'design' is a reflection of successive improvements on earlier prototypes – apes, monkeys,

Charles Darwin, born in 1809, was appointed naturalist aboard the *Beagle* survey ship in 1831. In the next five years he visited the coasts of South America, the Andes, the Galapagos Islands and Australasia, gathering data on their fauna, flora and geology. But his very controversial book *On the Origin of Species by Natural Selection* was not published until 1859, although the bones of his theory were contained in an essay written in 1844.

The little Galapagos finches shown here played a significant part in the elaboration of his theory. *Geospiza fortis*, the medium ground finch, shown above, has a robust, seed-cracking bill; its cousin on the left, *Geospiza scandens*, the cactus ground finch, has a bill adapted for manipulating cactus spines to dislodge insects from cracks. Both kinds of bill, Darwin reasoned, give their owners a survival advantage, in this case access to different foods.

other mammals, amphibians, fish, and so on. An organ such as the eye can be traced back through many simpler light-detection devices to the point where a few skin cells in a fish developed the ability to detect the change from light to dark or vice versa. Research in zoology and palaeontology has helped us to trace the evolution of each organ or body system, sometimes with great certainty, sometimes with a few gaps. But the overall trend is so striking that it is difficult to deny the basic hypothesis. The explanation of exactly *how* we evolved is less certain.

To understand something of how we have evolved can at least temper our natural reaction of awe at the complicated and specialized design of the human frame and organs. It would be an astonishing fact if the human heart, for example, had suddenly appeared with all its chambers, valves and muscles for the first time in the first human. But, as with the eye, we can trace back a line of earlier prototypes of the heart, to a creature that merely possessed a single tube with pulsing walls.

The word 'evolution' is more a description than an explanation. Nevertheless even this description is useful in reducing the magnitude of the explanation required. Instead of trying to discover how a fully functioning organ was so cleverly 'designed', we can ask the slightly easier but still extraordinarily difficult question: how did the successive improvements take place that led to the human eye, ear, heart, lung and so on? As no humans were around to observe these millions of tiny improvements, we can only make intelligent guesses, occasionally supported by circumstantial evidence. The person commonly credited with finding a coherent answer to the question was Charles Darwin, although other naturalists of his time had also started to reach similar rational explanations of our species' origins.

It is now generally believed that life forms evolve by a process called natural selection. Briefly, this suggests that each generation of a particular population of animals contains a few individuals that differ in some

Reticulated giraffes among the flat-topped acacias of Kenya's Amboseli National Park. Though not particularly well camouflaged to human eyes, their coat markings give them some immunity from their main predator, the lion. Their size is also of some survival value – lions very seldom attack full-grown giraffes.

small way from the rest. Most of these differences are harmful (what we would call deformities or handicaps) or neutral (perhaps a different eye colour or fur pattern) but, occasionally, individuals are born with useful characteristics that their brothers or sisters lack.

The usefulness of these characteristics lies in the fact that they give the individual some advantage in obtaining food, or a mate, or in avoiding danger – in other words they have survival value. These useful attributes are part of the animal's genetic make-up and so can be passed on to its offspring. So a characteristic that has survival value, even if it only produces a slight advantage, will allow more of the individuals that have it to live longer on average than those who do not have it. And because they live longer, they will be able to produce more offspring, and some of those will also have the improved characteristics.

An interesting aspect of this process is that really it is the environment that 'designs' the animal. Take, for example, the giraffe and its long neck. The suggestion is that among the giraffe's short-necked ancestors the occasional animal was born with a slightly longer neck. At a time when the foliage giraffes browsed on was becoming sparse on the lower branches of the trees but was more plentiful higher up, the animal with a slightly longer neck would have an advantage over its fellows: it would find food more easily and survive for longer. So the longer-necked animals would breed and produce some longer-necked offspring, who would have the same advantage. Over many generations, the longer-necked varieties would thrive and the shorter-necked ones starve and die out.

But none of this would happen if there were no trees with foliage out of reach of some of the giraffes. If foliage was equally available at all heights, the occasional individual with a slightly longer neck would have no advantage over its fellows. Indeed, it might even be at a disadvantage in some way: perhaps it would bang its head on low branches more often or be more conspicuous to predators. In any case, without the presence of taller trees and a shortage of foliage on smaller trees, longer-necked animals would not have an advantage over their fellows, and would not therefore come to dominate and eventually supplant the shorter-necked animals. In that sense, the presence of taller trees in the environment has 'produced' the giraffes.

For each of the millions of improvements that have resulted in the anatomy and physiology of the human body, there was an aspect of the environment that 'produced' it. Hard ground 'designed' claws and fingernails, and the speed of breakdown and absorption of food molecules 'designed' a long alimentary tract.

Almost everywhere you look you can see examples of 'design by environment', some more extreme than others. Here are two particularly striking products of environmental design: a saguaro cactus and a stick insect, both marvellously adapted to their environments.

The cactus *Cereus giganteus* is a native of Arizona; it alone survives where other plants have given up. Like all true cacti it has abandoned leaves in favour of spines, because leaves lose too much water – its green stems do the photosynthesizing instead. It flowers only when conditions are suitable for offspring. And its vertical furrows send what little dew and rain there is down to the roots as quickly and efficiently as possible.

The stick insect has been 'designed' not by climate but by predators. The adults of the species shown below, *Extatosoma tiaratus*, a native of Australia, fool predators by looking like dead twigs; immature individuals have leafy excrescences on their legs and bodies. The half-grown insect shown here perfectly mimics its leafy surroundings. And to cap it all, it keeps up a slow swaying motion!

Of course, the word environment is used here in a very wide sense. The environment of an organ deep inside the body may be the physics and chemistry of whatever enters and leaves it. For the human nervous system, the environment may be the physical qualities of light rays, or the influence of potential predators. But each step along the road of evolution by natural selection produces an advantage, and each advantage means that the environment can be coped with better and exploited further.

Every human organ or system shows the results of this shaping process, which has taken many millions of years and stages to reach its present form. The lung is a good example. It is 350 million years since one of our ancestors – the fish – achieved a tiny advantage in its watery world by swallowing air instead of oxygen dissolved in water. From that first minor change we have reached the stage where we obtain all the oxygen we need from the surrounding air, using a design that is perfectly adapted to the job. Buried in a person's chest is 50 square metres or so of extremely thin membrane across which oxygen and carbon dioxide continually change places even during the most extreme forms of physical activity. Surrounded by chest muscles to pump it in and out, the lung looks as if it has been designed for a modern human in the modern world.

The evolutionary process can only work if, as mentioned above, each new generation of individuals contains a few members with new characteristics. However many taller trees there are, for example, they will have no effect if short-necked animals only ever give birth to short-necked animals.

There is no doubt that populations of animals have this variability. With fast-breeding creatures such as insects, scientists have actually demonstrated evolution by natural selection by changing the environment under test conditions, and then seeing new generations appear that are better adapted to the new environment.

The conventional answer to the question 'how?' used to be to talk of genetic mutation. Every creature has half a blueprint for its descendants coded in its genes, as described in Chapters 23 and 24. This blueprint can occasionally be altered at random by radiation or chemical activity in our environment. It is as if, when thousands of copies of this book were printed, a gremlin in the printing press occasionally changed a letter or figure here or there. Most such changes would be detrimental – 'the lpving body' instead of 'the living body'; a few would be neutral – 'farther' instead of 'further'; and perhaps one or two would have a disproportionately strong effect – 'the loving body' instead of 'the living body'.

Such changes can actually be observed when genes are experimentally subjected to radiation. Some elements in the genetic code are changed and the resulting creature has some genetic material unlike that of either of its parents. The problem is that the positive effects of these 'mutations', as they are called, are usually very small. The negative effects, in contrast, can be dramatic. One change in a gene regulating just one chemical can be lethal. But the sort of changes that may be necessary to produce a valve in a blood vessel where no valve existed in the parents involves many genes, not just one element of one gene. This suggests that there must be a mechanism by which whole sections of genetic material change rather than individual elements. To take the book analogy one stage further (or farther), at the very least the gremlin must be able to rearrange some of the words on a page, or even tear out or insert whole pages, or even compose entirely new sections of the book!

The likely explanation seems to be that during the copying of genetic material from one generation to the next, some sections may be deleted, duplicated, swapped, transposed and so on, leading to minor or major changes, a few of which might be useful.

However they are caused – and there are many ingenious theories to account for them – these changes do occur, although very slowly. They are not usually observable in humans or other mammals because of the time it takes each species to be born, mature and reproduce. To produce the accumulation of genetic changes that shape the eye or the heart or the kidneys takes millions of generations, each one contributing a small element to their ultimate shape or structure.

Human evolution is a continuing process which began many millions of years ago. From the first living cell, 3500 million years ago, to generations alive today, there have probably been 1000 million handovers of genetic material, each one providing an opportunity for diversity to arise.

As we explore in this book the latest model in this long line of living bodies, we should think of each organ and system as having an ancestor, a predecessor that can be traced back and back through time. From a tube to a heart, from a swim-bladder to a lung and from a fin to an arm, the human body owes its present shape and structure to the environment it lives in.

One of the main factors that makes us unique is the way we use our body – in other words, our behaviour. Time and again in this book we shall see how the brain and the nervous system interact with the other organs and systems to produce conscious or unconscious human behaviour.

The living body is a world that is unknown to most people. Indeed, we cannot know about it without the help of the tools and techniques of modern science. In this book, we have tried to use some of those tools and techniques to provide images and explanations that will help to uncover some of the mysteries of how the living body works. When we do discover how each system and organ works, the mysteries may vanish, or they may be replaced by even stranger mysteries and a greater sense of awe. But this journey into the body will certainly not lead us to take our bodies for granted.

Four generations at the same table, and four handovers and reshufflings of literally millions of genes. In theory at least, genes are immortal; we die but they go on, hitch-hiking from one generation to the next. Was biologist Richard Dawkins, author of *The Selfish Gene*, right when he suggested that each new body is no more than a survival machine for the genes that made it? Because such large numbers of genes intermingle during sexual reproduction – some enhancing each other, some cancelling each other out – each new body is something of an experiment.

2 · Skin and Surface Senses

If you were asked which is the largest organ in the body, you might think first of the liver, the stomach or the lungs. But the correct answer would be the skin. This continuous tissue, which covers an area of nearly two square metres and weighs about four kilograms, *is* an organ in its own right; it has its own blood supply (about a third of the blood pumped from the heart goes to the skin) and its own specialized cells and glands. Although we may dismiss it as just a protective covering that stops our insides falling out, it has many other important jobs to do, and the deeper we look beneath the surface, the more complex it seems. One of its major functions – as the point of contact with the outside world – is to receive sensory messages from the external environment which are passed on to the brain.

The senses

The only link our brains have with the outside world is through the senses. Without them, it would probably be as if we were in a coma. Unable to receive sensory messages, we would remain oblivious of what was happening around us; whatever danger threatened, whatever chaos lay ahead, we would not be aware of any need to think about it or to use our brains and muscles to do something about it. Fortunately we have evolved ingenious ways to make sure that we *do* know what is going on outside ourselves – the intricate nerve network that makes up our five senses of sight, hearing, touch, taste and smell.

The combination of these enables us to put together a detailed picture of our world, a three-dimensional picture that stretches from the surface of our bodies to the furthest point we can see. In this chapter we will look at the three senses that build up a picture of the part of the world that is closest to us. Taste and touch, and even smell, rely on direct contact with parts of our environment. Sight and hearing, on the other hand, work by detecting light and sound given off by objects that are not in contact with us and could, in fact, be almost any distance away – the light we see from the sun, for example, travels about 150 million kilometres (93 million miles) before reaching us.

When a message arrives at the body surface, it first makes itself known by triggering some sort of receptor (a specialized nerve ending that has evolved to receive information and turn it into nerve impulses). There is a whole range of different types of receptor, and these are scattered throughout the body, ready to detect important signals inside and out. But once a receptor has been triggered, the sequence of events that follows is very similar, and does not depend on the type or location of the receptor.

Every receptor has links with a nerve fibre, a long cell that takes sensory information to the brain or some other part of the nervous system. In some cases – the skin of the toe, for example – the message is carried from the receptor to the lower spinal cord in one long, thin nerve cell a metre or more long: there it connects with nerve fibres leading to various parts of the brain.

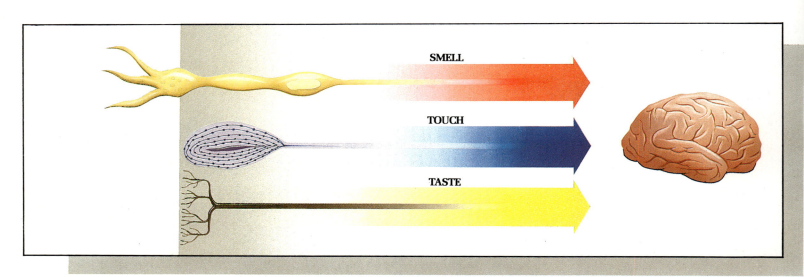

Our skin is our contact with the outside world, and three of our senses (shown in the diagram below left) are embedded in it, feeding information back to the brain. Smell depends on mucus-covered nerve endings that protrude just above the surface of the skin, or rather the specialized skin, of the nasal cavity; touch and taste are sensed by specialized nerve endings just below the surface.

Some of our sensory nerve endings are extremely sensitive indeed. Touch sensors in the fingertips can detect two stimuli only 3mm ($\frac{1}{10}$in) apart and a single smell receptor can probably detect one single molecule of certain substances. But continuous stimulation of any receptor causes it to adapt, lowering its sensitivity, though this returns after a brief respite. Our sweetness receptors can recover in a few seconds.

Smell

Not all sensory messages have to travel far. The receptors concerned with smell, for example, are very close to the brain. The sense of smell is a remarkable feat of chemistry: so sensitive are the sensory cells that only about half-a-dozen molecules arriving at a receptor cell are needed to trigger the cell to convey a smell message to the brain. And not only can our sense of smell detect minute amounts of chemicals in the air, it can also distinguish between about 3000 different odours, including some made artificially and so never encountered before. Although it is not known exactly how the detection system works, it may have something to do with the shape of the molecules. For us to be able to smell a substance, it must give off some of its constituent molecules so that they are floating in the air, and the molecules must be able to dissolve in the mucus that bathes the detector cells. Not all substances that evaporate readily have a smell – water, for example, is everywhere in the air but we do not smell it at all. When we inhale through the nose, the air gets dragged up past a small patch of tissue at the back of the nose. This patch contains some five million receptor cells, each with about half-a-dozen fine hairs coated with mucus, which trap and detect smell molecules.

When the molecules make contact with one of the hairs on the receptor cells, they cause the cell to send a sensory signal along a short nerve fibre that travels through a tunnel in the bone of the skull to the olfactory bulb, which is only a finger's breadth away. The olfactory bulb is an outcrop of the brain itself and from then on the message is treated like any other sensory message coming into the brain.

One clue to the way we identify different smells is that, in general, molecules that produce similar smells are similar in size and shape. One theory suggests that molecules of substances which have a particular smell fit into minute specific sockets on the surface of the detector hairs. These sockets would be shaped so that only one particular shape of molecule fits in each of them – rather like a lock-and-key mechanism. If this theory is correct, there must be thousands of different-shaped sockets, at least one for every different odour we can distinguish.

The 'lock-and-key' theory also helps to explain another interesting aspect of the sense of smell – the adaptation that occurs when we are exposed to a particular smell for any length of time. Sadly, the bouquet of a glass of wine lasts for only a few sniffs. From then on, the molecules are still in the air but we cannot smell them. Similarly, if we sleep in a bedroom with a vase of flowers, we stop noticing the smell after a

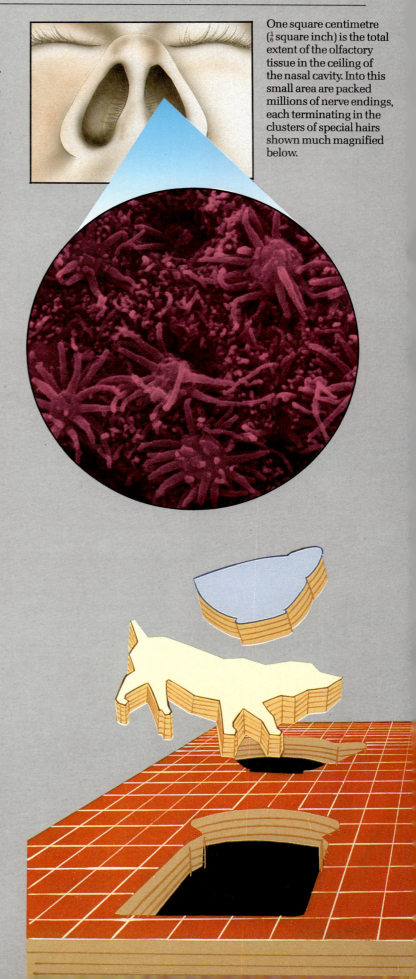

One square centimetre ($\frac{1}{6}$ square inch) is the total extent of the olfactory tissue in the ceiling of the nasal cavity. Into this small area are packed millions of nerve endings, each terminating in the clusters of special hairs shown much magnified below.

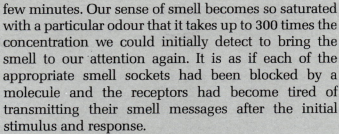

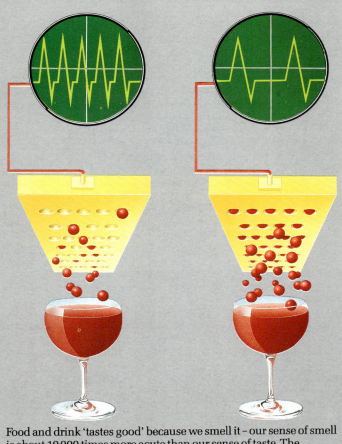

Food and drink 'tastes good' because we smell it – our sense of smell is about 10 000 times more acute than our sense of taste. The receptors in our nose respond faithfully to the first few sips of wine; then, as more and more sites on their hair-like endings are occupied by the volatile molecules of the wine, they lose their ability to send messages to the brain.

few minutes. Our sense of smell becomes so saturated with a particular odour that it takes up to 300 times the concentration we could initially detect to bring the smell to our attention again. It is as if each of the appropriate smell sockets had been blocked by a molecule and the receptors had become tired of transmitting their smell messages after the initial stimulus and response.

The usefulness of this 'blocking' arrangement, which applies to the other senses as well, is that we are aware of *changes* in our environment rather than having to cope with a constant stream of messages to the brain reporting that things are still much the same. Our immediate environment is full of smells that we no longer sense because they do not require our attention. But if another person comes into that environment from somewhere else, he will smell the odours that we have stopped noticing because his receptor sockets are not already occupied – for him there are *changes* in the sensory messages, whereas for us the messages are what we are accustomed to. If a new smell enters our environment, on the other hand, it will find unoccupied receptor sockets in our nose and come to our attention straight away.

For smell, as for most of the other senses, the sensitivity of the individual receptors is astonishing. The body has evolved to extract the maximum information from our environment so that we can form a comprehensive mental image of the changing world around us. In humans, the sense of smell is quite sensitive but it is not as important to us as it is to many animals. Dogs, for example, rely heavily on their sense of smell, which is correspondingly more acute than ours. Although, generally, the sensory tissue in a dog's nose and a human's is about the same size, a dog's has some 100 million receptor cells, each with more than 100 sensory hairs – compared with about five million cells, each with about six hairs, in humans. So perhaps it is not surprising that a dog could smell some substances at the extremely low concentration of only one gram if they were spread throughout the entire volume of air above a city the size of Birmingham or Boston, up to a height of 100 metres (320ft).

It is not known how many 'smell slots' we have. In reality we do not have a slot for bread, a slot for milk, and so on. But it is likely that we have a basic number of slots – as many as 30 according to some experts, and as few as 7 according to others – which respond to certain primary smells. Combinations of some or all of these give us different smell sensations.

Taste

Like smell, taste is chemical sense. But whereas we can smell *thousands* of different odours, we can detect only four basic tastes – sweet, salty, sour and bitter. We may think that a delicious and carefully prepared meal is a subtle mixture of numerous different tastes, but in fact the subtlety comes from the smells that add to the four taste sensations as we eat. This is why, when we have a cold and the patch of sensory tissue in the nose is blocked, our food seems much more tasteless. There are even nerve pathways in the brain that link the taste and smell areas, so that, perhaps, our memories for particular foods inextricably combine the two sensations.

The actual job of tasting is carried out by taste buds scattered over various parts of the mouth. There are some at the back of the throat and down into the larynx, although the most familiar are the groups of taste buds we can see as bumps on the surface of the tongue. Taste buds favour one of the four flavours – sweet detectors predominate near the front of the tongue, salty all round the edge, sour at the sides and bitter at the back.

Each taste bud contains up to 30 receptor cells, which are bathed by saliva containing whatever it is we are eating. Unlike smell, the mechanism of taste detection is probably not connected with the shape of the molecules, but may depend on some other chemical property – sour substances are usually acids, for example, whereas salty tastes are associated with molecules that break up into electrically charged components when dissolved in water. Unusually for sensory cells, those that detect taste are short-lived. Mature cells die after only a few days, and so the composition of a taste bud changes continually as new cells develop to replace those that have died.

Apart from the pleasure they provide, tastes have some protective value. For example, many of a group of substances called alkaloids have an unpleasant bitter taste, and many are also poisons – so the slightest taste can provide a memorable experience we will know to avoid in the future, if we survive.

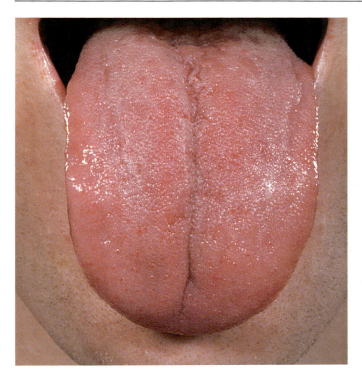

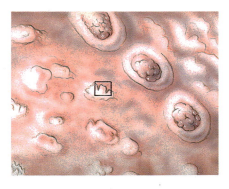

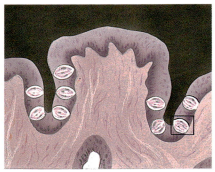

Food has to dissolve in the film of saliva on the tongue before we can taste it. Each of the small bumps or papillae on the tongue – the circular photograph below shows a much magnified papilla – is surrounded by a trench which fills with saliva and dissolved food. Opening into each trench through small pores (left and below) are the taste buds themselves, up to 200 of them.

The pasta meal opposite, sweet and salty, would stimulate the taste buds near the tip of the tongue.

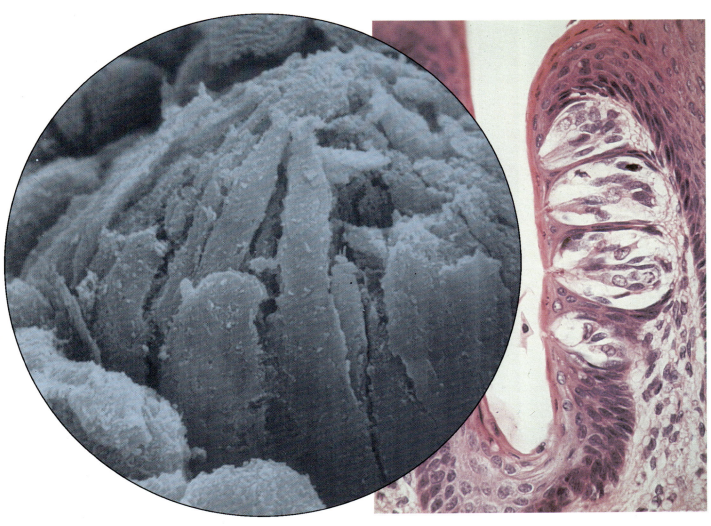

Touch

There are actually several touch senses, each of which is associated with a different type of receptor embedded at a different level in the skin. Some are sensitive to a light touch, others respond to pressure, others to temperature and yet others signal pain.

We need accurate messages from the skin to the brain. For safety's sake we need to have early warning devices in all parts of the surface of the body. Pressure, vibration, injury and temperature could all threaten our survival if they were too intense, and so there are nerve endings embedded in the skin that detect these factors in time for us to do something about them. For example, in any small area of skin there are receptors that register heat. If a hand is touching something hot, these receptors fire nerve impulses at a higher rate and then slow down, having signalled the change in the environment. Again, the receptors tend to respond to change rather than any absolute value, as described

earlier for the sense of smell, so after an initial burst of activity when a new touch sensation begins, our touch receptors settle down to a lower level of activity. Thus, unless we consciously monitor the sensation, we are not constantly aware of the feeling of the clothes on our back or the chair on which we are sitting, and so our awareness is released for more important things.

The receptors that respond to contact with the skin are not evenly distributed over its surface. Instead, they are clustered where they are most useful – in the fingertips, for example, we can discriminate between two simultaneous contacts only 2 or 3 millimetres (about $\frac{1}{10}$in) apart, while the skin of the back has far fewer receptors and the contacts can be several centimetres (an inch or more) apart before we feel them as separate.

If we look at one small area of skin, for instance the fingertip, we can see examples of all the types of touch receptors the body contains. Beneath the familiar surface ridges of the fingertips are different types of

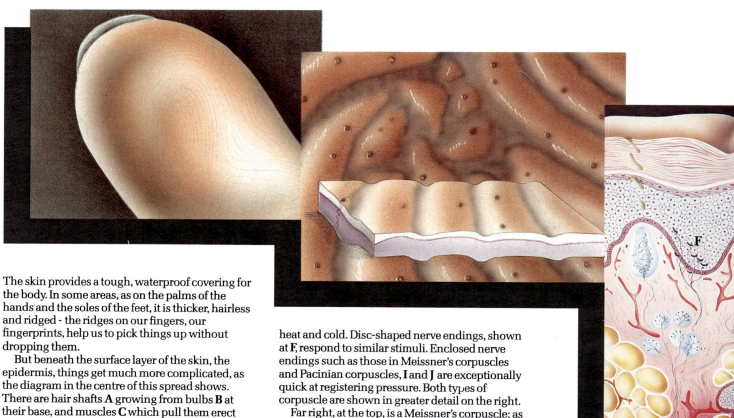

The skin provides a tough, waterproof covering for the body. In some areas, as on the palms of the hands and the soles of the feet, it is thicker, hairless and ridged - the ridges on our fingers, our fingerprints, help us to pick things up without dropping them.

But beneath the surface layer of the skin, the epidermis, things get much more complicated, as the diagram in the centre of this spread shows. There are hair shafts **A** growing from bulbs **B** at their base, and muscles **C** which pull them erect when it is cold. There are also glands – tightly coiled sweat glands **D** and sebaceous glands **E** which produce fatty substances to keep hair and skin waterproof. And there are many different kinds of nerve endings.

Free nerve endings such as those shown at **G** and **H** respond to gentle pressure and moderate

heat and cold. Disc-shaped nerve endings, shown at **F**, respond to similar stimuli. Enclosed nerve endings such as those in Meissner's corpuscles and Pacinian corpuscles, **I** and **J** are exceptionally quick at registering pressure. Both types of corpuscle are shown in greater detail on the right.

Far right, at the top, is a Meissner's corpuscle; as befits a structure specialized to respond to the slightest pressure it sits just beneath the epidermis. The Pacinian corpuscle below it is much bigger, responds to vibration and heavy pressure, and lives down among the fat cells **K** which house a large part of the body's long-term energy reserves.

nerve endings, which detect surface contacts. It seems that the different touch sensations are identified by direct physical effects on the different types of nerve endings. Some of these sensory endings are like trees, some have blobs on the end, and some are enclosed in a capsule, but no one type of ending has been associated with only one sensation. All of them send signals if they are bent or squashed in some way, and it seems to be where they are situated in the skin that determines what the sensation feels like.

Many of the signals that arrive at the surface of the body are associated with pleasure – gentle touching, tickling and warmth seem to be universally enjoyed. Then there are the mild warnings – it is getting colder, we are being squashed into a corner, something is too heavy to lift. But by far the most urgent messages that arrive at the skin are messages of pain. There do not seem to be any receptors particularly associated with pain, just lots of nerve endings in the skin that can carry

pain messages in certain circumstances, although any specializations of the endings may just not be apparent. Sometimes we feel pain when normal sensations become too intense – light, sound, touch and temperature can all cause pain at high levels. But not all overstimulation causes pain – you cannot produce pain by higher and higher concentrations of sugar on the tongue, for example. Also, some quite moderate types of stimulation, like a pin-prick or a small amount of salt in a wound, can produce an intensely painful sensation – it is likely that anything which damages tissue releases a chemical which triggers a pain signal in a nerve fibre.

Smooth skin, as on the fingertips, contains the variety of receptors that we have just seen. But hairy skin has only the bare nerve endings, sometimes lurking free just beneath the surface, sometimes wrapped round the base of a hair follicle. This is why pulling a hair can be sharply painful, because the movement is transmitted directly to the nerves at the hair's base.

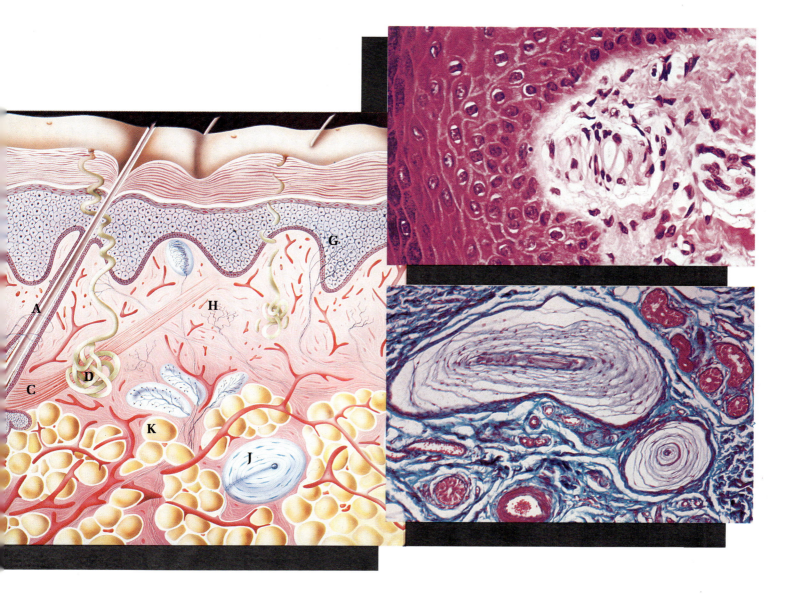

Other functions of the skin

Hair, like the skin from which it grows, is part of our protection from the outside world. Our ancestors were almost completely covered with hair but with us it survives only in strategically placed patches, although it is not always clear why hair is distributed as it is in modern man. On the head at least, the function of hair seems to be to protect us from various environmental hazards. The top of a bald head suffers all the year round. In summer it is at right angles to the rays of the sun, and the skin there could be easily damaged. In winter, it provides a large surface from which heat is lost. By retaining a layer of hair, or at least a hat, we avoid both these hazards.

As happens with all improvements in the living body, hazards in the environment lead to changes in the characteristics of our species. In a world in which, as far as we know, man evolved in the bright tropical sun, a skin that protected him from the harsh effects of ultraviolet light was obviously an advantage. And so specialized cells evolved which produced dark granules of pigment to protect the inner tissues from damage: melanin, abundant in black people, and carotene, a yellow pigment found in many Asian peoples, block the harmful rays. As the human race spread over the world, to colder, less sunny areas, their skin did not need protecting from the sun. In fact they often needed to seek out what sun there was. In addition, darker skin radiates away more heat than lighter skin – a positive disadvantage in cold northern areas where it might be

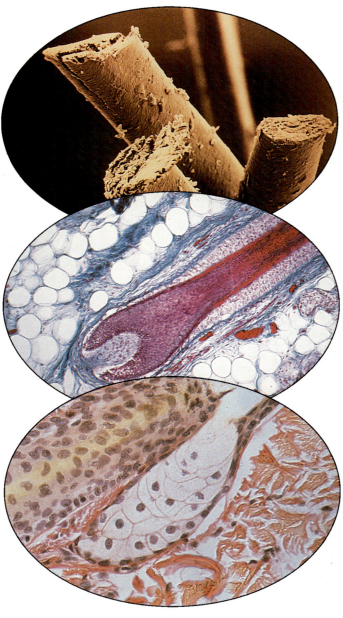

The top picture shows beard hairs after a shave. Beneath is a hair root, the living and growing part of hair. As cells are pushed up towards the skin surface they degenerate and die, leaving the tough outer layer that we lovingly comb and shampoo. Opening into the shafts occupied by the hair roots are sebaceous glands (bottom left); the sebum they produce is slightly bactericidal as well as waterproof.

vitally important to minimize heat loss. As a result of these factors, the peoples living in colder regions gradually evolved lighter-coloured skin. It could be a reminder of those days when we were all darker-skinned that those melanin-producing cells still become active when we expose ourselves to dangerous amounts of ultraviolet light – we get a suntan. The sun also helps the skin with one other important function – the production of vitamin D. This is produced in the skin when it is struck by ultraviolet light. If we do not get enough exposure to sunlight we become short of vitamin D, and need to take in more in our diet.

So the skin is more than merely a thin boundary between us and the world. It is a multi-layered organ with its own blood supply and nervous system. At its

deepest point, new cells are constantly being produced and they then push their way towards the surface, changing shape and function as they go. When they reach the top, they die, but even in death they perform a useful function: they form a tough protective layer to keep out germs. This protective layer is continually flaking off – in a lifetime, we will shed about 20 kilograms of dead skin cells.

A renewable waterproof coating, with variable protection for changing light and temperature, and the largest sensory organ in the body – the skin is all these and more. As we will see time and again as we explore the living body, nearly every organ and every system has evolved to perform more than one task, and the skin is a supreme example of that achievement.

Our ancestors in Africa had dark skins but in the course of colonizing cooler parts of the world their descendants lost their protective pigmentation. In the cross-section of skin below the melanin-producing cells appear brownish. The sun's ultraviolet rays stimulate them to produce melanin.

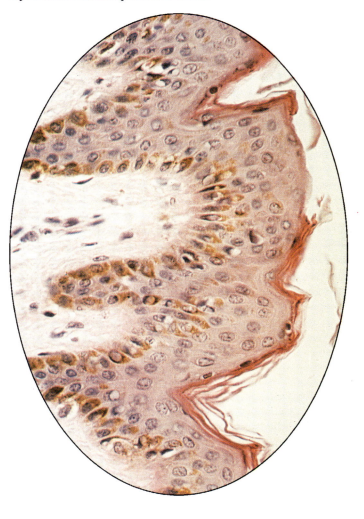

3 · Eyes and Ears

Of our five senses, sight and hearing are the most useful. They are our early warning systems, keeping us in touch with potentially hazardous events that have not yet reached the surface of the body, where touch, taste and smell can sense them. Much of what we know about the outside world comes to us in the form of light patterns or sound waves, detected by highly specialized sense organs. But detection is not enough. A pattern needs to be interpreted before it can be of any use to us, and only intimate collaboration between the brain and sense organs makes that interpretation possible.

Hearing
We are awash in an ocean of sound waves: from all directions ripples of sound arrive at our ears, carried as pressure waves in the air. Some form of pressure-wave measuring device developed in the skin of our marine ancestors and was used to detect underwater activity that might pose a threat, offer a partner or signal food. By the time man had evolved, the pressure-detector had become more complex and was buried at the end of a short tunnel in the head. And there were now two of them, which meant that even before we attempted to analyse the significance of the sound, we could discover the direction it had come from.

With two ears, we hear two sound messages, and only if the source of the sound is directly behind or in front of us do they arrive at our ears at exactly the same time. Even a slight movement of the source, or of the head, produces a difference in the arrival times of the sound at the two ears, and we use this time difference to work out the direction of the source. Remarkably, even a deviation of only a degree or so from directly ahead of (or behind) us can be detected by the time difference, which is then less than a ten-thousandth of a second.

In practice many sounds arrive simultaneously at our ears from different directions and at different intensities, and they all end up as a succession of simple movements of the eardrum. This is a taut disc of skin at the end of the tunnel that leads from the external ear, the visible part of the ear with which we are all familiar.

Between the loudest and the quietest sounds we can hear there is a hundred-trillionfold variation in intensity, and yet we have evolved a system which prevents the ear from being overloaded with loud noises but still allows it to detect quiet ones.

To help us detect quiet noises, the auditory canal (the passage leading into the ear) doubles the energy in sound waves by focusing them against the eardrum. Beyond the drum are three linked bones – the smallest bones in the body – that act as levers to magnify the sound vibrations further, doubling their effect again. The bones then transmit the sound waves to another vibrating membrane, the oval window, which has an area only about one thirtieth that of the eardrum and therefore magnifies the intensity of vibration by another 30 times.

These amplification mechanisms would, of course, be harmful with very loud sounds, and so there are also systems that have the opposite effect. Normally the sound is transmitted directly to the oval window by a bone called the stirrup. For very intense sounds the angle of contact between the stirrup and oval window changes so that the effect of the sound waves is reduced. There is a muscle that tightens up the three bones to make them more resistant to movement. And finally another muscle causes the eardrum to stiffen so that it too is less responsive to loud noises.

By the time the pressure waves arrive at the oval window, they have become confined to a much narrower range of intensities. Now the most useful information can be extracted from them. One of the most astonishing skills in the human body is the ability to separate different individual sounds from one complex sound wave. However many different sounds arrive at the ear, the eardrums and the inner ear membranes can perform only one action at a time, vibrating at a greater or lesser rate. A combination of sound waves is therefore reduced to a single long string of vibrations. But this one complex wave-trace still contains within it information about all the sounds, and each one can be separated out by the ear and the brain.

Any sound is a series of pressure waves in the air. The number of waves that pass any point in one second is known as the frequency of the sound, and any pure sound – a musical note, for example – has a constant frequency. The more waves per second, the higher the frequency, and the more high-pitched the note sounds to us. Beyond the oval window is a coiled, snail-shaped organ called the cochlea. Notes at different frequencies which arrive at the ear are transmitted through a fluid that fills the cochlea. The fluid surrounds a long, thin

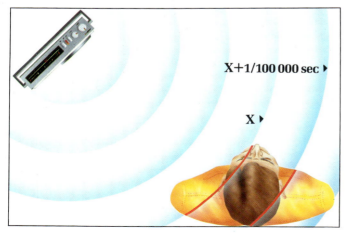

X+1/100 000 sec ▸

X ▸

Vision and hearing give us information about the *intensity*, the *frequency* and the *direction* of light and sound waves. In busy traffic, for example, a cyclist registers how far away other vehicles are by the intensity of the noise they make. He analyses the frequency of light to decide whether a traffic light says stop or go. And he uses the direction-finding capabilities of his eyes and ears to build up a three-dimensional model of the surrounding traffic in his brain.

    Much of our information about the three dimensions of our world comes from pairs of organs working together. One way in which the brain works out the direction of a sound source is by detecting the difference in arrival times of the sound waves at either ear (left). Even if the difference is only one hundred thousandth of a second the brain will judge the direction of the sound accurately.

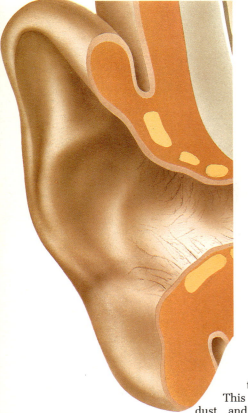

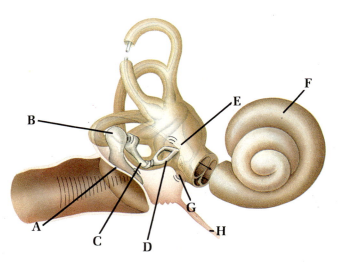

The drawing on the left shows the system of vibrating membranes, little bones and fluid-filled chambers that enable us to hear and keep our balance.

A is the eardrum. **B** is the ossicle, the hammer, to which the eardrum is attached; **C** and **D** are the second and third ossicles, the anvil and the stirrup. **E** is the oval window and **F** the cochlea. **G** is the round window and **H** the tube which leads to the nasal pharynx to equalize air pressure on either side of the eardrum.

Below is a much magnified view of the ossicles; the stirrup is in the foreground on the left.

The ear flap acts as a collecting trumpet for sound waves, and funnels them into the ear passage. This is lined with hairs that trap dust and other invaders, and with wax-secreting glands similar in appearance to sebaceous glands. In healthy ears wax continually moves to the outside, carrying with it dead skin cells and other debris.

membrane that runs the length of the cochlea. If unrolled, this membrane would be wider at one end, near the outer ear, and taper to a point about three-and-a-half centimetres away. It has two rows of cells, called hair cells, which are moved back and forth by the sound waves in the fluid. They are almost like the keys and strings in a piano: just as each key and its string correspond to a particular note, with the high notes at one end of the piano gradually progressing to low notes at the other end, so the hair cells in a particular region of the membrane respond most readily to a particular sound frequency. Although the pattern of response on the membrane has low notes at one end and high notes at the other, the response is more complex, because every hair cell is stimulated to a certain extent by every frequency.

With a high note, a wave travels from the oval window along the membrane, coming to a peak fairly soon and then dying away towards the tip. With a low note, the travelling wave builds up more slowly and peaks much nearer the narrow end of the membrane. With both the high and the low notes, all cells are moved by the wave, but any one cell moves a different amount, depending on the frequency of the sound.

Each cell sends its signal down a nerve fibre; the whole bundle of fibres forms the auditory nerve, which travels to the part of the brain, the acoustic cortex, that is specialized to interpret signals from the ear. Here, the same frequency layout that we saw in the cochlea is repeated – a strip of brain tissue that reproduces the piano keyboard, with the brain cells that deal with high notes at one end and low notes at the other. It seems that not only is there a frequency map in the brain, but also an auditory space map, so that single cells in the auditory cortex respond not only to frequency but also to position in space. Sound intensity is signalled by a combination of increases in the frequency of nerve impulses and increased numbers of fibres signalling.

There are more than 20 000 hair cells in the cochlea and each can have as many as 100 hairs protruding from it. These are so sensitive that they can detect a movement at the eardrum of less than the diameter of a hydrogen atom (which is about one quarter of a hundred-millionth of a millimetre). Or, if the eardrum were a kilometre wide, the hairs could detect it bending by less than a centimetre.

Loudness is measured in decibels – a whisper is about 20 dB, a jet plane taking off about 140 dB. Loudness becomes increasingly painful, and harmful, above 100 dB. It takes about 36 hours to regain normal hearing sensitivity after $1\frac{1}{2}$-2 hours of continuous loud disco music.

There could be problems with having such sensitive devices buried in our heads, where there is a continual source of vibration. Every time we speak, sound waves boom through the cavities in the mouth and nose and reverberate in the skull. Every time we walk, our muscles and bones set up low-frequency vibrations that could deafen us. But we are not deafened, probably because we tighten up the muscles that damp down vibrations in the ear bones. Our hearing is most sensitive to a frequency of about 3000 Hertz (vibrations per second), although there is no obvious reason why. This frequency is considerably higher than that of the normal human voice, so this extra sensitivity provides no real advantage in hearing and understanding each other. It has been suggested, however, that the reason for our hearing being most sensitive to this particular frequency is because it happens to be the principal pitch of a human scream.

Sight

The variations of intensity and frequency of sound waves in our environment are matched by similar variations in light waves – their brightness and colour. With light, however, our senses have an additional source of information they can use, based on the fact that light travels in straight lines. We can get a general sense of direction from sound waves, but we cannot form well-defined 'images' from them, although blind people develop much better skills in this respect. But with light we have developed a sense that can present to the brain very accurate models of the outside world in terms of brightness, colour and shape.

Like the ear, the eye has evolved ways of reducing extremes of intensity. Light first travels through the cornea, the transparent outer covering, which bends it to help focus images further back in the eye. It then passes through the pupil, a variable-sized hole in the iris, which is wide open in conditions where we have to make the most of the little light that is available and which closes down to a pinpoint when we are in very bright light. Extra amounts of pigments in the iris – if you have brown eyes, for example – provide extra protection against the distracting effects of very bright light, although blue eyes see better in less light.

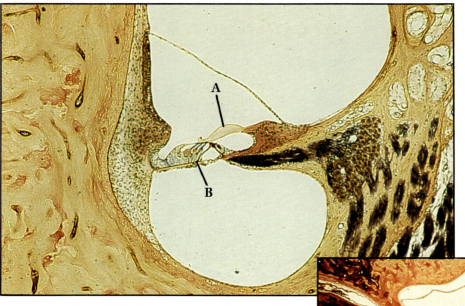

The diagram on the right shows the principle of the cochlea. Vibrations in the fluid of the top canal cause vibrations in the vestibular membrance **C**, which vibrates the fluid in the smaller canal beneath it, which in turn vibrates the tectorial membrane **A**, which brushes the tips of the hair cells (shown here as a graduated row of little tuning forks, which send signals to the brain for analysis.

The auditory nerve is shown at the bottom of this page, on the right.

Above is a cross-section through one coil of the cochlea. The organ of Corti, the 'keyboard' that responds to pitch and loudness, is in the middle. As the tectorial membrane **A** vibrates, it tickles the hair cells on the basilar membrane **B** into action.

On the right is a slice through the three coils of the cochlea; six cross-sections of it are therefore visible. Note the thick bundle of nerve fibres that leads to the brain.

The photograph below, taken with the aid of a scanning electron microscope, shows the astonishing symmetry and beauty of the hair cell keyboard. The hair cells are arranged in rows and each has its own crescent of fine hairs tuned to respond to certain frequencies and intensities.

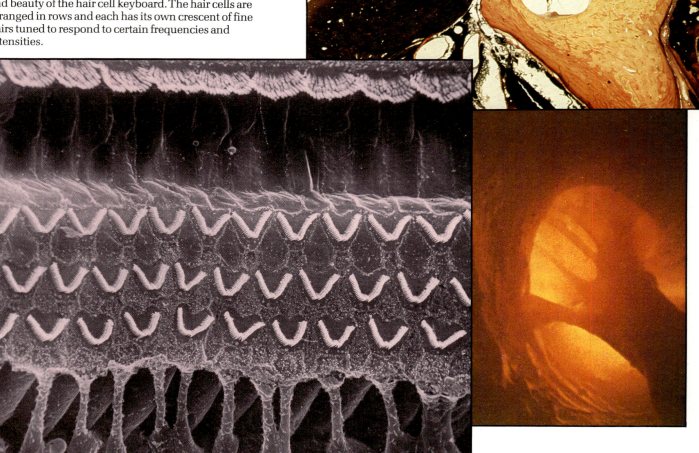

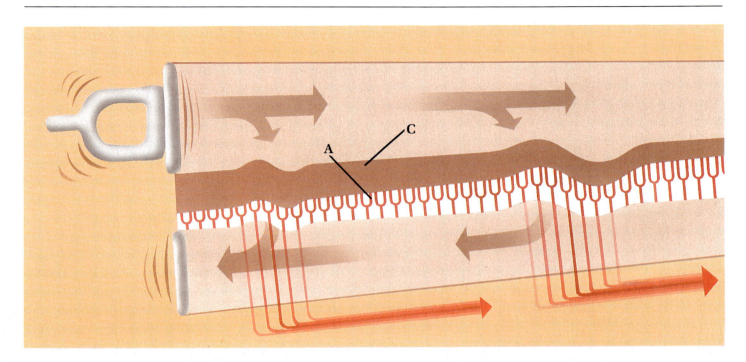

Beyond the iris is a transparent lens that transmits light waves to the back of the eye, the retina, and performs the finer focusing of the incoming light (already partly focused by the cornea) so that some of the light rays form a sharp image on the retina. The lens achieves this by changing shape to bring objects at different distances into focus. When we want to see nearby objects most sharply, the muscles surrounding the lens contract to let it expand into a near-spherical shape; when we want to direct our attention to distant objects, the muscles relax and pull the lens thinner to bring the image into focus on the retina.

The retina, coating the inner curved surface at the back of the eye, is one of the most remarkable pieces of tissue in the human body. Paper-thin and 6 square centimetres (1 square inch) in area, in each eye, this layer receives the whole rich imagery of the human world – colour, movement, depth, light and shade – and provides an immense amount of information for the brain to interpret. It is easy to forget quite how small an area this image covers. When we look out into the world around us, we are aware of a 'big picture' that stretches from just behind the left shoulder to just behind the right, and from above the forehead to below the waist. Every detail seems pin-sharp, colours are full of subtlety, some objects stand in front of others, and the whole scene appears solid. And yet the actual image that this is based on is the size of a large postage stamp, projected onto a constantly moving piece of tissue, as the eye moves to survey the world, and the only part of the image that is really seen in detail is the size of a pin-head.

Like a newspaper photograph, the retina breaks the scene into dots – 125 million of them. These correspond to rods and cones, specialized detector cells in the retina that respond to the brightness and colour of the light that falls on them. The retina is actually several layers thick, because the light detectors need nerve cells to carry their messages away and a blood supply to keep them nourished.

If you were designing a surface to receive images in this way, you would probably make sure that the light detector cells had an uninterrupted view of the light rays. In fact, in the human eye, light has to travel through the layers of blood vessels and nerve cells in the retina before it reaches the rods and cones. However, the brain manages to ignore these permanent obstructions and pays attention only to the temporary variations in light and colour that reach the rods and cones. This is another example of how useful it is when our senses detect change rather than absolute amounts of a stimulus.

The two types of receptors have very different functions. The rods are found over most of the retina and act as simple brightness detectors. If a single unit of light, a photon, hits a rod, a chemical reaction occurs which bleaches a pigment in the rod for a fraction of a second. As soon as the bleaching has occurred, a series of chemical reactions restores the pigment to its former opaque state, ready to respond to another photon of light. The rate at which this bleaching and regeneration takes place is responsible for an electrical sensory message that travels away from the rod to the optic nerve. The rods are sensitive to quite low levels of light

The structure of the eye:

A cornea
B iris
C lens
D lens muscles
E sclera (the tough outer
tunic of the eyeball)
F choroid layer (contains
lots of pigment and
blood vessels)
G retina (light-sensitive
layer)
H optic nerve

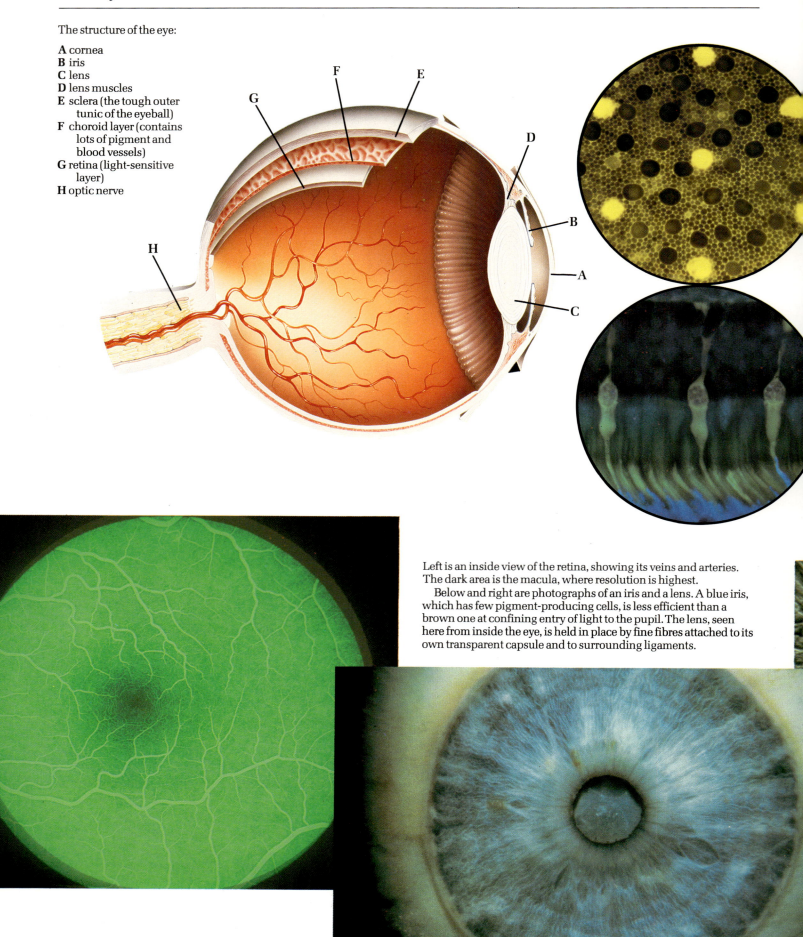

Left is an inside view of the retina, showing its veins and arteries. The dark area is the macula, where resolution is highest.

Below and right are photographs of an iris and a lens. A blue iris, which has few pigment-producing cells, is less efficient than a brown one at confining entry of light to the pupil. The lens, seen here from inside the eye, is held in place by fine fibres attached to its own transparent capsule and to surrounding ligaments.

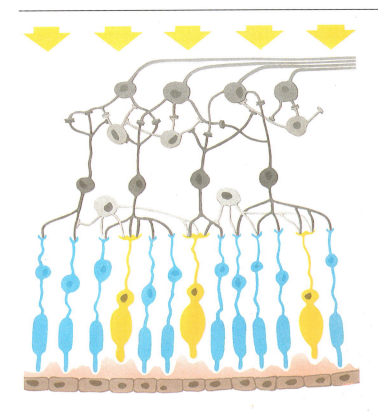

The diagram above shows the various layers of nerve fibres and nerve cell bodies that intervene between the rods and the cones and incoming light. The photograph on the left shows what this arrangement looks like in real life – the larger cells are the cones.

Away from the macula, the central point of the retina where acuity and colour perception are greatest, small, tightly packed cones give way to a mixture of rods and large cones (the yellow dots in the photograph above left are cones); near the edge of the retina there are up to 30 times more rods than cones.

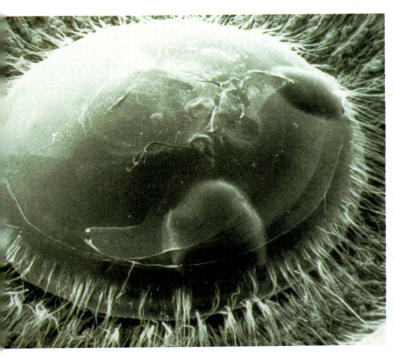

but they are unable to detect different colours. This means that where there is very little light we see things in black and white, or shades of grey.

An interesting difference between us and some animals is that our eyes have evolved more for perceiving sharp detail than for the ability to see at night. We therefore lack the reflective membrane at the back of the eye that makes cats' eyes so dramatic at night. This membrane has the effect of doubling the amount of light hitting the detector cells of the retina, so making the eyes more sensitive. But it also makes the details of the image much more blurred. Our eyes have a dark absorbent coating at the back of the eye, which means we cannot see so well in the dark but do not suffer from the blurring effects of the reflected light.

Cones, the other group of light-sensitive cells, are most concentrated near the centre of the retina. They have a different pigment from that in rods and, instead of responding only to brightness, they react to three different colours – red, blue and green – in different ways. Some cones respond best to red light, others to green and a third group to blue light. By some processing method in the brain or the nerve cells of the retina (an extension of the brain), we combine the individual messages from the different colour receptors and interpret the correct colour of what we are seeing. In this way we see pure white as one colour when, in fact, it really results from the three different colour receptors being stimulated equally; this process is similar to the way a white image on a colour television screen is actually made up of red, green and blue dots shining with equal intensity.

As well as enabling us to see colour, the cones are responsible for our extremely good visual acuity – the ability to discriminate fine detail. Someone with normal vision can just see a small coin at a distance of 65 metres (210ft), provided the image falls in the centre of the retina. It must have been a great advantage to the first animals to develop accurate vision. Man's ability to spot danger signs minutes or even hours before the threat arrived had considerable survival value in helping him to avoid unpleasant surprises.

Because the cones are packed very densely in the centre of the retina – 150 000 or so per square millimetre – we can see detail far more clearly if the image is focused on this area. Our eyes have the reflex ability to bring any important part of the visual scene to this part of the retina for detailed examination. This visual acuity is enhanced by the fact that the cones are extremely narrow in the central part of the retina, one fifth or less of the diameter of the cones elsewhere in the retina, and so the centre of the image can be broken down into smaller

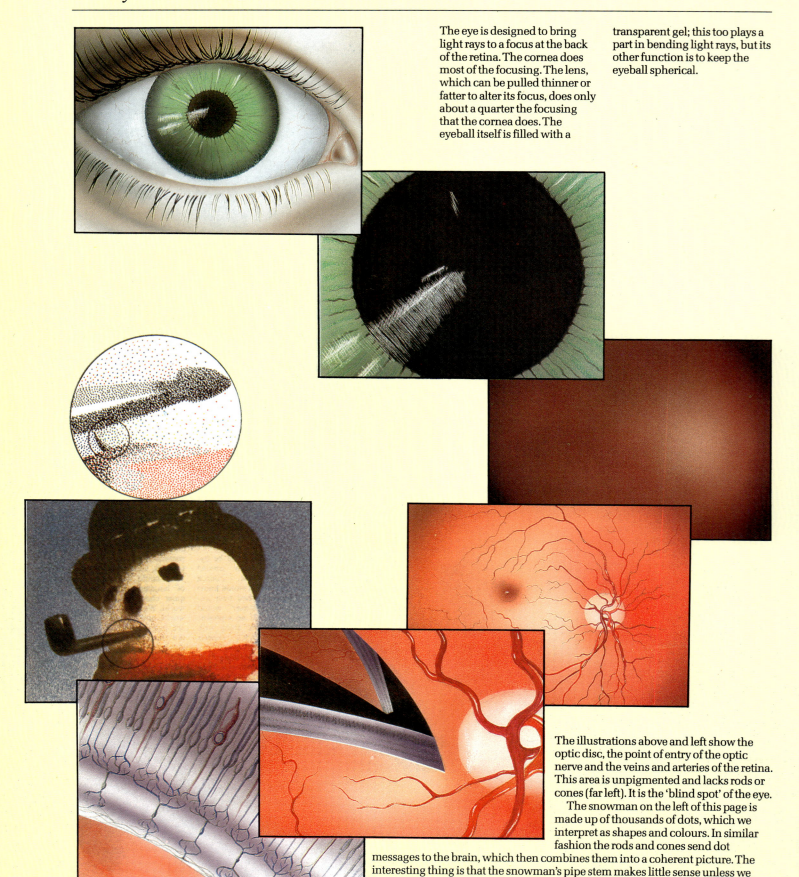

The eye is designed to bring light rays to a focus at the back of the retina. The cornea does most of the focusing. The lens, which can be pulled thinner or fatter to alter its focus, does only about a quarter the focusing that the cornea does. The eyeball itself is filled with a transparent gel; this too plays a part in bending light rays, but its other function is to keep the eyeball spherical.

The illustrations above and left show the optic disc, the point of entry of the optic nerve and the veins and arteries of the retina. This area is unpigmented and lacks rods or cones (far left). It is the 'blind spot' of the eye.

The snowman on the left of this page is made up of thousands of dots, which we interpret as shapes and colours. In similar fashion the rods and cones send dot messages to the brain, which then combines them into a coherent picture. The interesting thing is that the snowman's pipe stem makes little sense unless we have seen the picture it is part of. We are programmed to see and recognize wholes, not details.

elements for the nervous system to analyse.

Our retinal image tells us the shape and colour of an object in the outside world, but it does not really give us a very good idea of its distance. And yet it is obviously important to know whether a visible threat is near and small or far away and huge.

The first creature to acquire two eyes that looked at the same scene gained a great advantage in the way it saw the world. Because the eyes are a few centimetres apart, they each see any scene from a slightly different angle, and the brain compares the pictures from the two eyes. With a near object there are large differences between the two retinal images. The outlines of the object may be similar but the backgrounds are likely to be very different, because of the different direction in which each eye is pointing. If the object is further away, the backgrounds are much more alike. By looking at a scene with near and distant objects in it, you can observe the differences in the two images by opening and closing each eye alternately.

These differences between the two images received

by the eyes are only one method of judging depth and distance in the world. There are messages to the brain from the muscles that move the eyes, which tell us where the eyes are pointing at any one moment. Since our eyes converge much more sharply on near objects than on distant ones, the brain interprets these messages to work out exactly where the two eyes are pointing at any time. Even with one eye, we can use several elements in the visual scene to judge distance. Overlapping contours, perspective, haze in the atmosphere – all of these combine with information from two eyes to build a vivid three-dimensional image.

As we will see in Chapter 9, the gathering of light and the detection of brightness and colour are just the first stage in supplying us with an accurate picture of the world. As with sound, it is the brain that really makes sense of the flood of sensory messages, helping us to organize properties like shape, colour, pitch and intensity into the perception of the physical objects that fill the space outside our bodies.

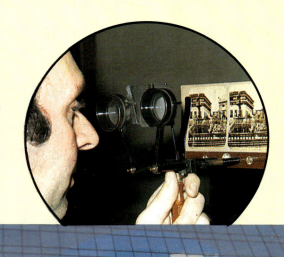

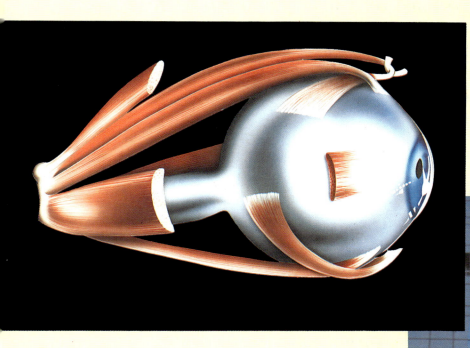

Six muscles move the eyeball in its socket – two on either side, one above, one below, and two wrapped obliquely round the eyeball. They all attach to the sclera or tough outer jacket of the eyeball. The seventh muscle, shown cut short at the top left of the illustration above, moves the upper eyelid.

We rotate our eyes, and turn our head, so that we can look directly at scenes or objects, but only the middle of them is in focus; the rest is blurred, as in the photograph on the right.

The stereoscopic viewer above right is doing what our brain does so efficiently – superimpose two slightly different images to give a 3-D single image.

4 · Eating

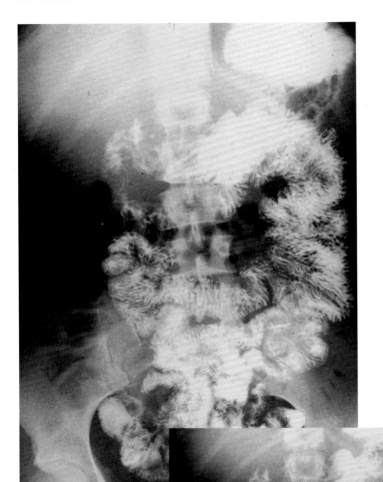

Hunger, like sex, is one of the body's strongest drives. Indeed, the two have a lot in common: they can both be pleasurable and both are essential for our survival. The sex drive leads to couples producing children and so the human race survives. Equally important, none of us would be fit enough to reproduce if we did not eat once or twice a day.

In order to grow, think and move, our bodies need regular supplies of carbohydrates, proteins, fats, vitamins, and minerals such as potassium and sodium, and we obtain them from our food. Some of these chemicals (carbohydrates, for instance) are 'fuel' for burning in the body, and others (such as proteins and vitamins) are needed to replace losses through wear and tear, excretion or chemical reactions.

All the food we eat travels along a tube through the body, from the mouth to the anus. At every stage of this journey there are specialized cells and tissues that help to extract the maximum value from the very wide range of substances we eat, and to store or get rid of what we do not need or cannot digest. In this way, through the collaboration of muscles, nerves, glands and blood, a

Three X-ray photographs of a barium meal passing through the gut. In the picture above, taken 15 minutes after eating, the stomach is well outlined at the top right; the two pictures on the right show the progress of food one hour and two hours after eating.

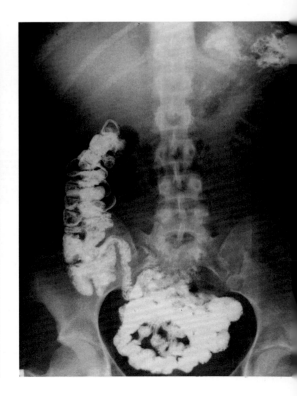

plate of normal food is converted into the daily requirements of the average human body.

In a typical day the human body needs to take in the following: just under 0.5 kilogram (1lb) of carbo-hydrates and fats, to supply the energy we need; about 60 grams (2oz) of protein, to provide the building blocks for our cells; 0.5 gram of calcium, for our bones and teeth; one hundredth of a gram of iron, a vital com-ponent of blood; and tiny amounts of various vitamins and minerals that are essential for our bodily processes but which we cannot make for ourselves – we need only a few thousandths (or in some cases a few millionths) of a gram of each of them every day, but would suffer and eventually die if we went without them.

Different foods contain different proportions of all these substances and so we have developed a sophisticated system for extracting and breaking down the ingredients we need and dealing with the remainder. The main element in the body's system for processing food is the digestive tract. We think of it as being well and truly inside the body, but in fact, since there are holes at both ends, the space in the tube is continuous with the outside; so in this respect we resemble an elongated doughnut. Somehow or other, food in the digestive tract (the hole in the doughnut) has to get through the walls of the tract and into the interior of the body. Before that can happen, we have to start things off by eating.

What makes us eat ?

There are two main reasons why we seek and eat food – hunger and pleasure. If we go without eating for long enough, we begin to feel hungry. This happens when we have used up significant amounts of important body components and their levels start to fall.

Deep in the brain there are two pairs of nerve centres which control our eating behaviour. One of them, the feeding centre, sends messages that stimulate us to eat ; the other centre seems to act only when our hunger has been satisfied, by damping down the feeding centre. Nobody knows exactly what leads to the feeling of hunger, but it is thought to occur when the cells in the body begin to find that there is not enough fuel available for their essential activities, and they have to slow down the rate at which they work. The satisfaction centre probably relies on nerve messages telling it of the presence of food in the stomach to make us stop eating. Between them, these two centres exert a fine control over food intake; during a lifetime, a person eats 20 or 30 tonnes of food but usually manages to maintain the same weight to within a few per cent.

Hunger is not the only thing that drives us to eat. Man has learnt to derive a lot of pleasure from eating, by experiencing the tastes and smells of the rich variety of substances that we eat. Even if we are not starving, we often eat some foods because it is a pleasurable experience to do so. These pleasant sensations play an

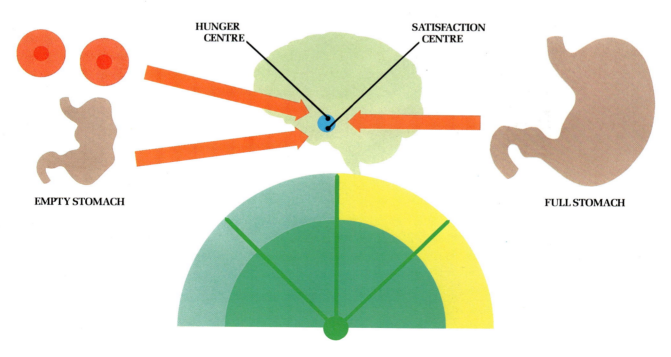

HUNGER CENTRE   SATISFACTION CENTRE

EMPTY STOMACH   FULL STOMACH

Dropping sugar levels in our cells, and the movements of an empty stomach, trigger a hunger centre in the hypothalamus of the brain which tells us it is time to eat. The sensation of a full stomach triggers another centre, the satisfaction centre, also in the hypothalamus, which tells us to stop eating.

The first phase of the digestion process usually starts before food even enters the mouth – seeing or smelling food causes us to produce saliva. Once food enters the mouth, nerve messages (the blue arrows on the central diagram below) are sent to the brain, stimulating it to instruct the salivary glands to go into full production.

Saliva, seen as bubbles in the photograph below right, has several different functions. It contains an enzyme that begins the starch-breakdown process; it dissolves food so that it can be detected by the taste buds; and it makes food slippery so that the tongue can manipulate it easily.

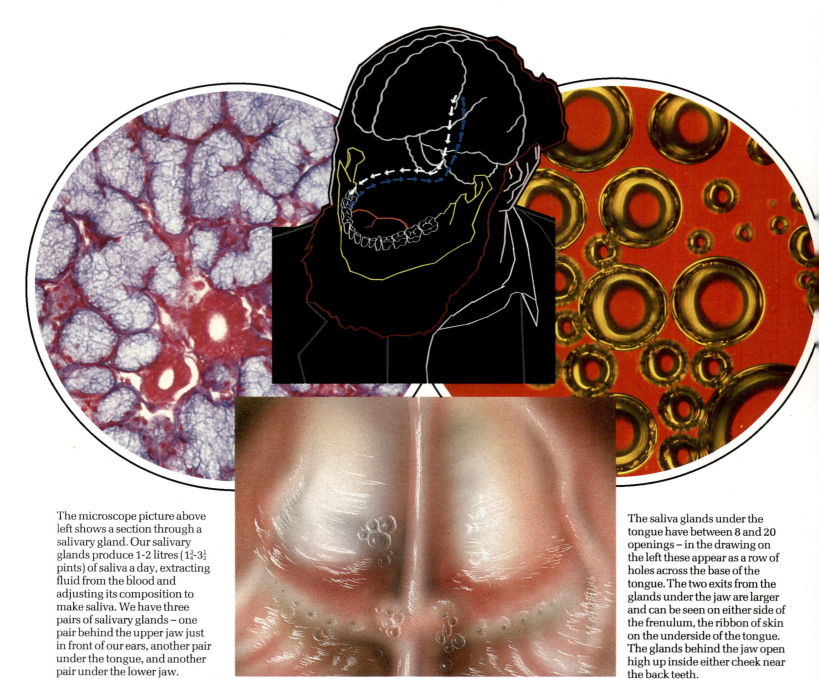

The microscope picture above left shows a section through a salivary gland. Our salivary glands produce 1-2 litres ($1\frac{3}{4}$-$3\frac{1}{2}$ pints) of saliva a day, extracting fluid from the blood and adjusting its composition to make saliva. We have three pairs of salivary glands – one pair behind the upper jaw just in front of our ears, another pair under the tongue, and another pair under the lower jaw.

The saliva glands under the tongue have between 8 and 20 openings – in the drawing on the left these appear as a row of holes across the base of the tongue. The two exits from the glands under the jaw are larger and can be seen on either side of the frenulum, the ribbon of skin on the underside of the tongue. The glands behind the jaw open high up inside either cheek near the back teeth.

important part in getting food into the mouth. They also have an important role in preparing the stomach and digestive system to process the food before it has actually arrived there.

The mouth

Even before the food reaches the mouth, the sight and smell of it can lead the digestive system to prepare to receive it. In the mouth, the salivary glands can start working so actively that a small sharp pain sometimes occurs as the specialized cells step up their activity, extracting from the blood a mixture of water and chemicals that goes to make saliva.

Every day we secrete about 1 litre ($1\frac{3}{4}$ pints) of saliva, which performs several useful tasks. It can kill bacteria, for example (which may be why animals lick their wounds). Part of the saliva is mucus, which protects the inside of the mouth from sharp particles and makes it easier for food to travel towards the back of the mouth and down the throat. The saliva also includes powerful enzymes – chemicals that start breaking down the carbohydrates in our food into the sugars that we can assimilate much more easily.

Saliva also carries molecules of food into the pits on the surface of the tongue and mouth that contain the taste buds. As the food churns around in the mouth and is broken up, we are able to derive the maximum benefit from its tastes and smells.

During the early stages of eating, appetite actually increases as we get our first exposure to the sensations of food. That is why we feel hungriest as we start to eat, and it is also probably why restaurants serve *hors d'oeuvres* – instead of filling us up too soon, as you might expect, they actually increase our appetite for the main course. And our taste buds can lead us astray later in the meal; when they become satiated with a particular range of tastes, we stop eating after a while, but if we experience new tastes near the end of a meal – with the dessert trolley, say – our appetite is roused again and we may end up eating a great deal more than we really need.

As we eat, the thorough mixing of the food is helped by the powerful actions of the teeth and jaws. The teeth have to break down any large chunks so that they are small enough to slip down the throat, and they can exert tremendous pressure between them with the help of the jaw muscles which, for their size, are probably the strongest muscles in the body. This pressure is fine for a piece of meat but would not be very pleasant if applied to our own flesh – and in the middle of every mouthful is a tongue, pushing and turning, twisting the food about and even talking. Fortunately, our brain constantly monitors the position of the tongue and coordinates the actions of the jaws and the tongue to prevent us biting our tongue or the inside of our cheek – or at least most of the time.

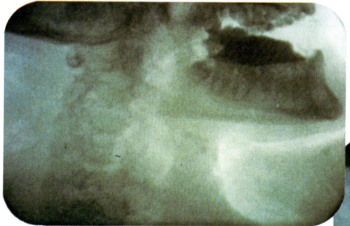

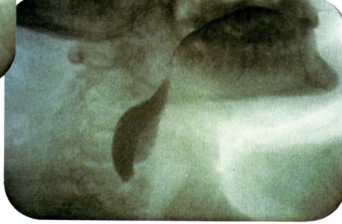

X-ray photographs, above and right, of a bolus of food in the mouth and as it slips into the oesophagus. The windpipe, or trachea, lies in front of the oesophagus, hence the pushing back of the bolus as the entrance to the windpipe is closed by the action of swallowing.

When the food is ready to be swallowed, a whole sequence of actions takes place with little or no voluntary control. These begin with the tongue pushing the food to the back of the mouth, sliding it towards the tube that leads downwards. Because of the economy of design of the body there is a certain amount of sharing among the tubes and passages in this part of the head. Nose, mouth, throat, oesophagus and windpipe are all open to each other for much of the time. We can breathe through nose and mouth, for example, and indeed swallow air. But when we are eating we do not want food to go up our noses or into our lungs and so the events of each swallow must make sure that only the appropriate passages are open and the others are closed.

Swallowing starts with the tongue closing off the front part of the mouth and pushing a ball of food (called a bolus) towards the back. As soon as the bolus nears the back of the throat, a trapdoor leading up to the nasal passages closes. You can feel this happening if you try to breathe and swallow at the same time. For a moment, no air can come down past the back of the throat, because the food is there. Similarly, as the food moves further down towards the oesophagus, the tube that will carry it to the stomach, another trapdoor covers the entrance to the windpipe. If food or liquid strays that way, there is an immediate attempt to get rid of it by coughing, which is just a means of pushing a short gust of air from the lungs to sweep the passage clean.

Once the bolus has been steered past all inappropriate passageways and into the top of the oesophagus, another type of reflex takes over. The oesophagus is not just an inert tube that allows the food to drop from mouth to stomach: it actually pushes food to the stomach with pulsing waves of muscle activity.

This reflex action is so effective that it even works against gravity, so you can swallow effectively even while standing on your head.

The stomach

Man eats a wider range of foods than almost any other animal, and yet all these foods, with their extremes of texture, temperature, acidity and composition, will eventually have to be broken down into a relatively small number of different types of molecules – mainly sugars, amino acids, fatty acids and minerals. Begun in the mouth, this process really starts in earnest in the stomach, with an intensive mechanical breakdown which reduces small chunks of food to a uniform mush. It is here that protein breakdown begins, caused by another enzyme. The stomach also acts as a store for the broken-down food and gradually lets it through to the intestine, where most of the chemical breakdown occurs and where the process of digestion is completed.

The stomach is not situated where most people think it is – at the level of the navel – but lies slightly below and to the left of the breastbone. The stomach is quite small when empty, with a capacity of half a litre, but can expand to hold up to five litres in order to accommodate large meals. It is able to do this because its wall consists of three layers of muscles which relax.

Liquids pass through the stomach fairly quickly, but solid foods have to stay there longer to be broken down into small enough fragments to get through the stomach's narrow exit. Experiments suggest that the solid food is arranged in layers in the order of being eaten, with the food eaten first being nearest the stomach wall where the important chemicals are secreted to start the process of digestion.

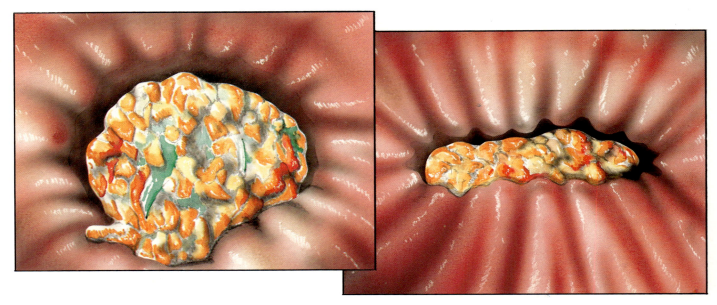

The oesophagus is one of the narrower parts of the digestive tract. It is about 35cm (14in) long and passes through the diaphragm, the sheet of muscle at the base of the lungs, in order to reach the stomach. The upper two-thirds of it have an extra layer of voluntary muscle as well as an outer layer of longitudinal smooth muscle and an inner layer of circular smooth muscle.

These contract rhythmically, powerfully squeezing chewed food (at the bottom of the opposite page) down towards the stomach. Food halts momentarily at the lower end of the oesophagus, hence the unpleasantly 'full' feeling when we gobble our food. Then the muscles at the entrance to the stomach relax, allowing the food to pass into it.

A glass of water with a meal is not so much an aid to digestion as a replacement for body fluids secreted into the digestive tract in order to carry out the process of digestion.

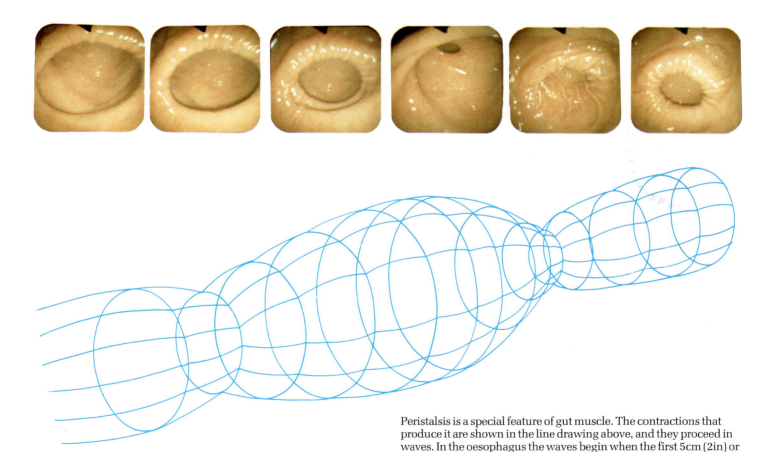

Peristalsis is a special feature of gut muscle. The contractions that produce it are shown in the line drawing above, and they proceed in waves. In the oesophagus the waves begin when the first 5cm (2in) or so contract; they then begin travelling down towards the stomach at a speed of up to 5cm (2in) a second. The time-lapse endoscope photographs above show the progress of a wave of contraction moving past the same point in the oesophagus.

The stomach secretes gastric juices in response to nervous signals from the brain but also, as shown in the diagram below, specifically in response to the hormone gastrin. Gastrin is produced in part of the stomach wall as soon as food comes into contact with it. The hormone then makes its way, through the bloodstream, to other parts of the stomach wall stimulating them to produce more hydrochloric acid and more enzymes to break down proteins and fats.

Below left is a scanning electron micrograph of part of the stomach wall. The white strands of mucus are part of the stomach's protection against corrosion by the strong acid it produces.

Below right is a computer model of the structure of pepsin. This enzyme digests proteins. It does not digest the protein wall of the stomach itself because it does not become active until it comes into contact with hydrochloric acid.

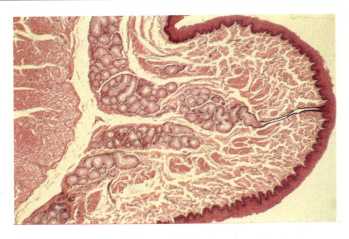

The oesophagus does more than just squeeze food down to the stomach; it also secretes mucus to help it on its way. Above is a section through the wall of the oesophagus in which the glands that produce the mucus can be seen very clearly; they have a distinctive coiled appearance. On the right is one of the ducts that empty the mucus into the oesophagus.

GASTRIN-PRODUCING CELLS

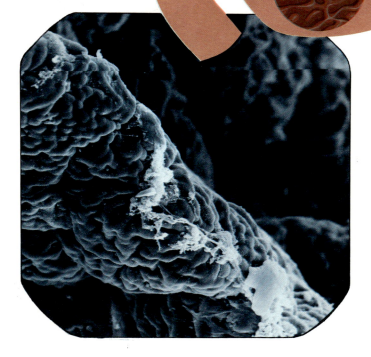

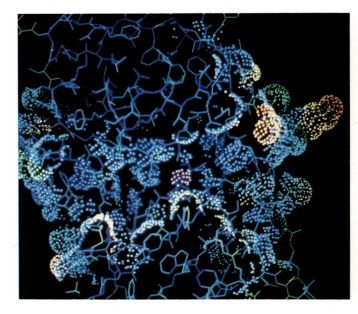

In the stomach wall there are cells that produce various chemicals that will be useful in digestion further down the intestine. One of these chemicals is hydrochloric acid – so corrosive that it can quickly damage unprotected cells and fibres, like the meat and vegetables we eat. It can burn a hole in human skin, and since our stomachs are frequently coated with the acid, it is fortunate that the stomach lining has a protective coating of jelly-like mucus which prevents damage. The coating is able to do three things simultaneously: keep acid away, let through some food molecules to reach the lining where they are absorbed, and let through stomach secretions in the opposite direction.

Unfortunately, although the stomach is protected most of the time, there are occasions when the acid can penetrate the stomach's protective coating, producing an ulcer. This can happen when chemicals like alcohol or aspirin interfere with the protective coating and allow the acid to attack. Ulcers can also occur as a result of an upset in the natural balance of gastric juices.

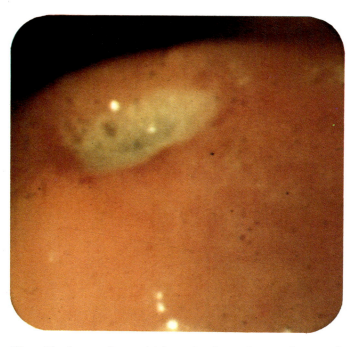

Ulcers like the one above, which can develop in days, can be caused by over-secretion of acid in response to stress or to a diet that is over-rich or over-refined. The ulcer in the picture is healing – fortunately the lining of the stomach repairs itself very quickly. The depression marks the spot where the lining of the stomach was eroded away.

The secretion of juices into the stomach and the way the stomach breaks down the food are controlled in a number of ways. There is a link between the nervous system and the stomach, for example, and our emotional state can affect the stomach. When someone is excited or angry, the stomach becomes overactive and its lining appears scarlet and swollen. Conversely, when someone is depressed or frightened, the activity of the stomach slows down, and its lining, normally a healthy pink with full blood vessels, becomes pale and dry. So it is perhaps not surprising that periods of strong emotional activity can eventually cause ulcers.

Similarly, merely the thought or expectation of food can start the gastric juices flowing – even before we have seen the food or started eating (up to about 30 per cent of gastric juice is produced in this way). When we eat, the juices are used to break down food, and once the food arrives in the stomach, gastric juices flow even faster. This is because of another element in the control of digestion – a hormone called gastrin. This is released from one part of the stomach when it comes in contact with food fragments. The hormone is released not into the stomach but into the bloodstream, which transports it to another part of the stomach where it stimulates the release of more gastric juices.

While all this is happening, the stomach is not just a passive bag. The top section is steadily squeezing food down towards the exit where vigorous waves of contraction squirt some of the food into the next part of the intestine. These contractions occur about three times a minute during and after a meal until the stomach is empty, which may not be for several hours.

The rate at which the stomach empties depends on how much we have eaten and how solid the food is. A full stomach empties faster, liquids go through quite quickly, and fats tend to float to the top of the stomach so that they linger longer before moving on. Sooner or later, however, all the food from our last meal will leave the stomach to be broken down further and absorbed in the intestine. The now empty stomach is much less active, apart from occasional bursts of contractions every 90 minutes or so which pass down the whole intestine. If we go long enough without food, we may even feel these contractions as hunger pangs and be stimulated to start eating again. But even if we have forgotten the last meal, it will be with us for several hours yet, moving slowly down the next seven or eight metres of intestine. The real function of all this eating is only just beginning – the absorption of the food to supply fuel and building blocks for the activities of everyday life.

5 · Breakdown

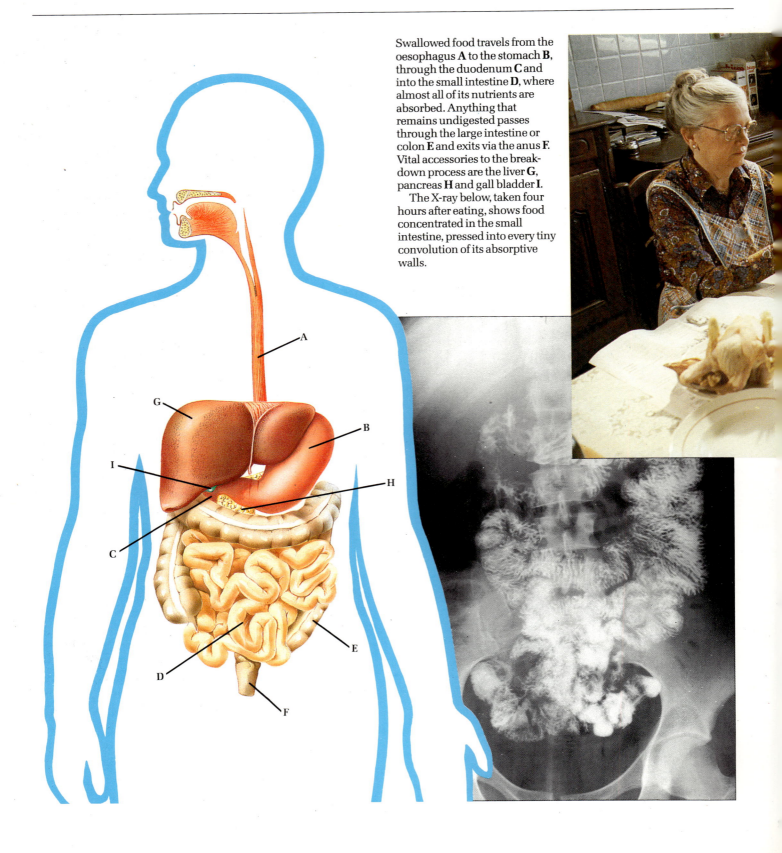

Swallowed food travels from the oesophagus **A** to the stomach **B**, through the duodenum **C** and into the small intestine **D**, where almost all of its nutrients are absorbed. Anything that remains undigested passes through the large intestine or colon **E** and exits via the anus **F**. Vital accessories to the breakdown process are the liver **G**, pancreas **H** and gall bladder **I**.

The X-ray below, taken four hours after eating, shows food concentrated in the small intestine, pressed into every tiny convolution of its absorptive walls.

Just as we prepare food by chopping and slicing, so chemicals called enzymes (the red arrows in the diagrams below) prepare food for absorption by chopping proteins into smaller units called amino acids, fats into fatty acids and complex sugars into simple ones.

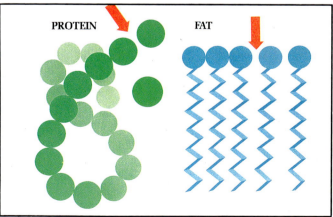

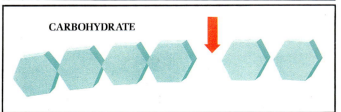

The rituals and ceremonies that accompany our meals are an important part of our social life. We do not plan our meals according to a detailed shopping list of substances that our bodies need – so much carbon, so much nitrogen, so many minerals; we eat in a way which reflects national and personal characteristics as much as nutritional needs.

But the real purpose of the whole occasion is to provide the body with certain substances. We transform our food into the pure and simple elements that our bodies need, and expel what is left after the body has performed its complex chemical processing. Without regular supplies of specific ingredients, we would soon be unable to renew the changing structures of our body. We would also be deprived of energy to carry out the day-to-day tasks on which our survival depends.

All our human energy comes originally from the sun. The plants and animals we eat are just a form of packaging for that energy. Plants can absorb solar energy directly but animals cannot. We take the plants, or animals that have fed on them, into our bodies by eating them, and then crack open their molecules to release the solar energy.

So there are two jobs that have to be done before we can make the best use of our food: breaking down the food in the alimentary tract, and then absorbing it and distributing it to where it is needed inside the body. These two tasks of breakdown and absorption can take place at the same time, during the 7-metre (23-ft) journey from the stomach down the first part of the intestine, known as the small intestine. They involve a close collaboration between several different parts of the body, including the bloodstream, the nervous system and hormones, and two organs of great importance for digestion – the pancreas and the liver. These are linked to the intestine by tubes and passages which carry a whole range of digestive juices, each with a specific target in the mixed-up mess of food.

Breakdown of food

We have to break down our food into its components because many of the molecules are too big to pass through the walls of the intestine or in the wrong form to be transported by one of several different methods used by the gut wall. The three most bulky constituents are protein, carbohydrates and fats, and each of them has a structure that has to be thoroughly broken down before we can use it.

Protein, which is found in meat, beans, cheese and bread, is in the form of long chain molecules. We need the individual links in the chain to build up our own different long chains, tailor-made for our own tissues.

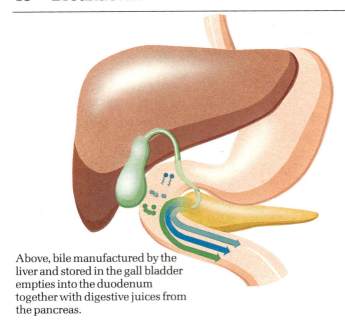

Above, bile manufactured by the liver and stored in the gall bladder empties into the duodenum together with digestive juices from the pancreas.

Above is an endoscopic view of the inside of the duodenum and a much magnified view of the tongue-shaped 'villi' that project from its walls.

The body therefore has to chop up the chains, with enzymes that attack each link, like chemical knives. Some of these individual links, or amino acids, are found only in certain foods, mainly animal protein, and it is essential to include these in the diet. Most of the others can be manufactured by the body if it needs more of one than another.

Carbohydrates, found in leaves, fruits, grains, potatoes and in very small amounts in some animal products, have large molecules that are really collections of sugars. We use each sugar in a slightly different way, although they all provide a source of quick energy for our physical activities. Here again, the task for our bodies is to break down the connections between the chains of sugars so that the individual components can be absorbed.

With fats the problem is different. Like the oil on top of vinegar in a salad dressing, the fats and the rest of the food will not mix without some help. The fats will need to be emulsified into small droplets that can then pass through the walls of the intestine to where they can be picked up by the blood and taken to where they can be stored as a long-term source of energy.

In addition, we need small amounts of vitamins and minerals which we cannot manufacture ourselves and are vital for many of our internal processes. These are not broken down but carefully split off from the food containing them.

Man has discovered how to start this breakdown process outside the body, by cooking food, using heat and sometimes water. In discovering how to make our food easier to digest we have also created a greatly increased menu of tastes and smells to add to our enjoyment of the necessary activity of eating.

As we saw in the last chapter, when the food reaches the stomach, it is well on the way to becoming a mush of identical-looking components, although in fact many of the different molecules are still waiting to be broken down. The important business of digestion has already begun with protein in the stomach and continues when the food starts to leave the stomach, through a narrow ring of muscle leading to the intestine. This muscular opening can control the rate at which food leaves the stomach and enters the duodenum, the first part of the small intestine. This is a curved section of intestine about 25 centimetres (10in) long which takes a small amount of food at a time from the stomach and starts to modify it with chemicals.

Some digestive juices are already waiting in the duodenum before the food arrives there, as a result of the sight and smell of food. But the main trigger for increased activity is the arrival of the food, which initiates several different processes. Groups of cells in the gut wall secrete a range of powerful breakdown chemicals, which will be mixed with the food, by the muscular squeezes of the intestine, as it passes down the next four metres or so.

Other chemicals are secreted by the pancreas and the gall bladder. The pancreas is a tadpole-shaped organ containing cells which produce a rich mixture of enzymes – chemicals that each attack very specific parts of our food and break them down. There is one that snips the joins along the length of the protein molecules, another breaks the carbohydrates into sugars and yet others break up fats. Food that reaches the duodenum contains acid from the stomach and sugars from the preliminary breakdown of carbohydrates, and these two substances trigger messages to the pancreas. When the messages arrive, the pancreas is ready to secrete a litre (1¾ pints) or more of its rich mixture through a thin tube, the pancreatic duct, that opens into the small intestine.

This exit also carries another important secretion – bile. This green fluid is a mixture of acids, produced in specialized cells in the liver. It flows down a network of tubes, meeting more and more outflows from other similar cells until it arrives in the gall bladder, where it is stored and concentrated as bile.

When fat arrives in the duodenum, the gall bladder secretes its contents, which help to break down the large drops of fat into much smaller droplets. Like washing-up liquid, the bile has a detergent action. Since many small droplets have a much larger surface area than one large one, the fat is now much more easily attacked and broken down by some of the enzymes that have come down the same tube from the pancreas.

Bile has a strong green colour, and this is the end result of the breakdown of red blood cells. The average red blood cell lives for about four months and is then destroyed. Its pigment is broken up, and some of it arrives in the gall bladder to form a dramatic colouring for the bile. When the acids are summoned to the duodenum, the pigments go too. They then travel the rest of the way down the intestine and supply the brown colour to the faeces.

Although the human diet varies widely from nation to nation and even from person to person, we all have the same biochemistry. By the time the rich variety of human food has been broken down, it has been turned into the same basic components.

Absorption

Unlike some animals, most of us do not eat all the time, but restrict our eating to regular periods intermingled

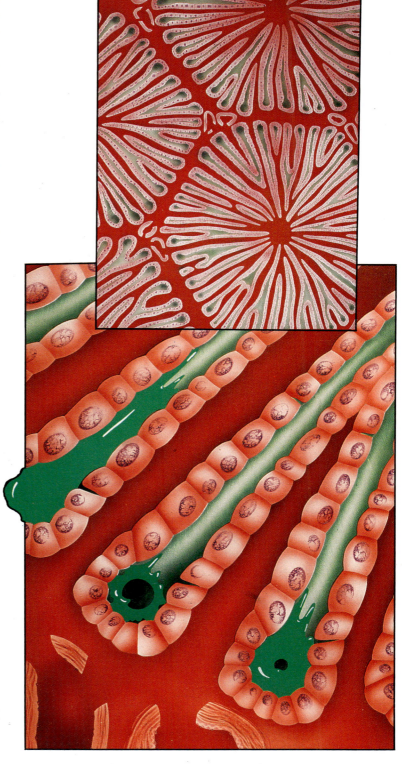

Liver cells are grouped into hundreds of thousands of polygonal lobules, each drained by a central vein and receiving food-rich blood from the intestines. Liver cells, in addition to their many other vital functions, produce bile, emptying it not into the bloodstream but into special channels (shown green above) which eventually join up and go to the gall bladder. Bile is rich in sodium carbonate.

with all our other activities. But the body needs the products of our digestion continually. There has to be a means of smoothing out the peaks and troughs of digestion and absorption so that our supplies keep coming through to the cells that need them. To cope with sudden demands for physical activity, we need to store the broken-down carbohydrates and fats somewhere until they are needed. To be prepared for the unexpected need to make new cells or proteins in a hurry, we have to store the amino acids where we can get at them quickly.

Part of the supply problem is solved by the fact that the food has such a long way to travel through the intestine. After leaving the stomach, the food has to travel the length of the small intestine and although the speed can vary, depending on what we have eaten and the efficiency of the gut's squeezing movements, it will take several hours. Absorption can take place for most of the length of this tube, and the walls have evolved to provide a huge surface area for the individual sugars, fats and amino acids to penetrate. Each square centimetre of the walls of the small intestine contains hundreds of finger-like projections which increase the

surface area by about 600 times, which is roughly the size of a doubles tennis court.

These projections wave around in the food that surrounds them, moved by their own smooth muscle. This helps the absorption of the food components, which travel through the surface layer and are carried away to the rest of the body. Some components are carried off by the blood stream and others, the fats, are transported by the separate circulation of the lymph system, carrying a fluid rather like blood plasma without the red blood cells.

Some of the components go straight to where they are needed, distributed throughout all the cells of the body by the blood circulation. Many substances are carried to the liver, where they are processed by a special system of blood vessels that links the intestine and the liver directly.

Weighing about 1.5 kilograms ($3\frac{1}{3}$lb), the liver is the largest single organ in the body and has innumerable different functions, perhaps running into hundreds. It is an energy store for the glucose that we get from carbohydrates in our diet and may not need to use straight away; it manufactures a number of important

The large intestine, displayed below, is about three times as wide as the small intestine and nearly 5m (15 ft) shorter. Undigested food is admitted into it, very small amounts at a time, through a valve. **A** marks the appendix.

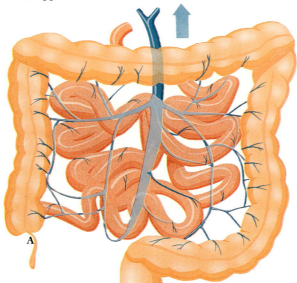

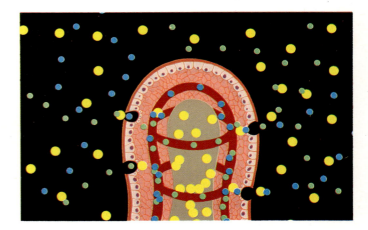

Less than 10 per cent of sugars, proteins and fats escape absorption by the villi, the finger-like projections that give the small intestine its amazing surface area. Simple sugars and amino acids (the blue and green dots in the diagram above right) are absorbed directly into the capillaries; fat (yellow dots), either as droplets or as fatty acids and glycerol, passes into the lacteals, tiny ducts which are part of the lymph system (see page 170).

The photomicrograph opposite shows the incredible number of blood vessels that supply the villi – the largest have taken up an injection of black dye.

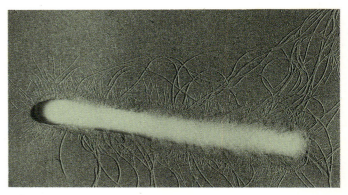

Above is a typical gut bacterium. Unlike the stomach, which is too acidic, the small intestine is quite a comfortable environment for bacteria, most of which are benign. Intestinal fluids can contain up to a million bacteria per millilitre.

substances such as blood-clotting proteins; it is a supply depot for iron and vitamins, manufacturing, storing and releasing them as necessary; it produces bile, which, as we have seen, is crucial for breaking down fats; and it produces heat as a by-product of its tremendous manufacturing activity.

Following a meal, the liver receives most of the end products of carbohydrate breakdown. The potatoes, bread and sugar that were eaten a few hours ago are now all broken down into the similar much smaller units of sugars like glucose. Most of these units are packaged into a more durable form of storage, called glycogen, which can be quickly converted back to glucose when it is needed for energy and for maintaining our vital blood glucose levels to which the brain is very sensitive. Failure to regulate the levels of blood glucose is a symptom of diabetes, and can be very serious if not correctly managed.

The fats in our diet, which are now in tiny droplets, are carried by the blood to special adipose cells for storage. These are the cells that many of us try to get rid of by diet or exercise, but meanwhile they wait in soft and rounded clumps defining the shape of the human body, particularly in women. These cells give up their fats to be broken down into energy supplies whenever necessary, especially during physical activity.

This journey through the gut may give the impression that there is a slow, stately progress of molecules through the body until they arrive at their ultimate destinations, but in fact the situation is more chaotic. Identical atoms of food can follow very different paths, some of them passing into the body and being used up in a few minutes, others staying in the body much longer, according to how quickly they happen to be used. At any one time there is a turmoil of chemical activity in which the components of many different meals are combined and recombined in our bodies, broken down for energy, reconstituted for storage, incorporated into our cells, and then released. The components of our food vary in the rate at which they turn over in the body. A carbon atom from a potato may stay in the body for only a few minutes before being breathed out as carbon dioxide. A nitrogen atom from a piece of steak may spend several years doing duty as part of a protein molecule in the wall of a brain cell, or perhaps a muscle cell.

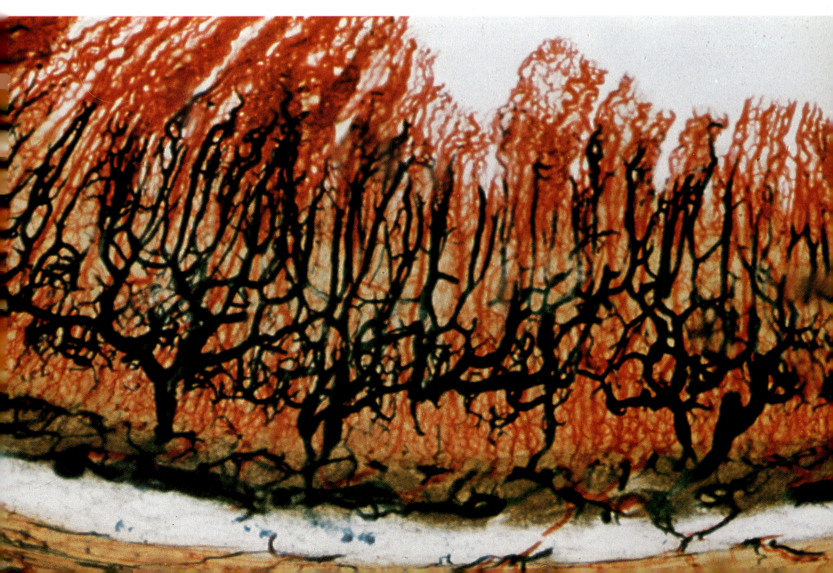

The cells in the liver, the hepatocytes, are among the most versatile in the body, each one carrying out many different jobs simultaneously. Under high magnification the cells of the liver appear roughly cuboid. Threading between them are tiny tributaries of the arteries and veins, and tributaries of the duct that eventually delivers bile to the gall bladder.

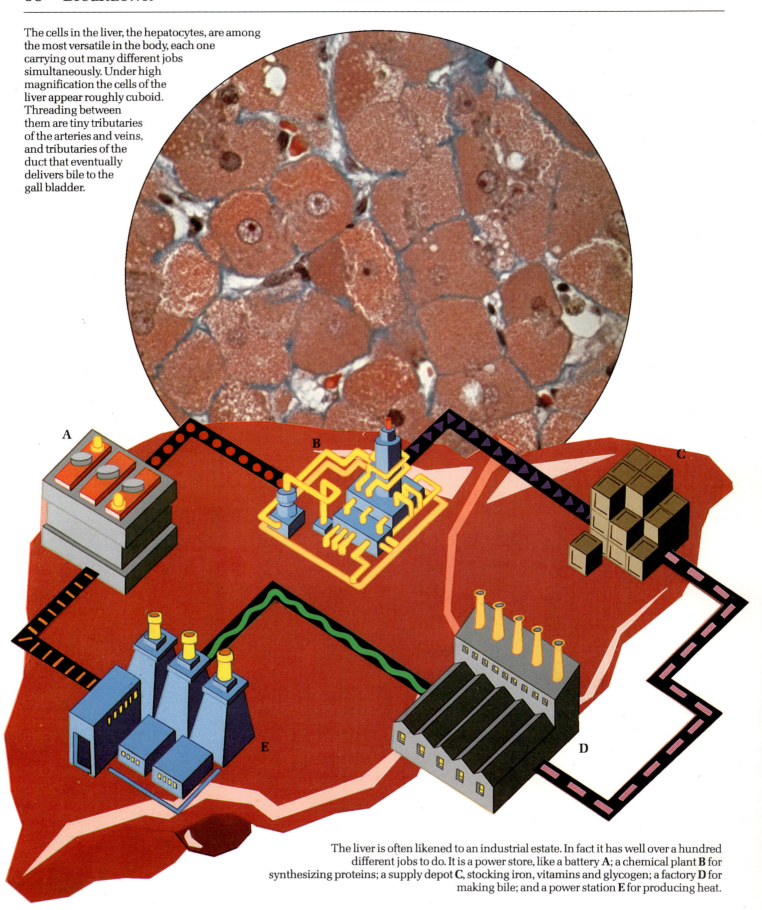

The liver is often likened to an industrial estate. In fact it has well over a hundred different jobs to do. It is a power store, like a battery **A**; a chemical plant **B** for synthesizing proteins; a supply depot **C**, stocking iron, vitamins and glycogen; a factory **D** for making bile; and a power station **E** for producing heat.

The large intestine

While lunch is being eaten, breakfast has arrived at the final stage of its journey, and the part that takes the longest time: the large intestine. A much more straightforward tube than the small intestine, its 1.5 metres (5ft) has to bend a few times to fit into the average abdomen, but what is left of the food has a much less tortuous journey.

Near the beginning of the large intestine is the apparently useless and sometimes troublesome blind alley known as the appendix. Probably a store for plant-eating bacteria in our ancestors in their herbivorous days, it is now a shadow of its former size and can become infected by food that has been deposited there by mistake and has become stagnant. However, this comparatively rare disaster has not been enough of a problem over the millennia to lead to the evolution of people without an appendix.

From this point onwards, the residue of our food that we do not need is compressed and solidified. When it leaves the small intestine for the colon – the main part of the large intestine – the residue is liquid. This has helped with movement through the gut and with the absorption process but, so that we do not lose more water from the body than is necessary, the waste will now become dried and more solid as some of the water is reclaimed.

Much of what is left consists of the cell walls of the plants we eat, made of indigestible cellulose. The more of this fibre there is in our diet, the more bulk there is in the waste, which therefore travels faster through the large intestine, as waves of pressure in the walls push the bulky residue along. With a less fibrous diet, various problems, both major and minor, tend to occur in the colon. The waste travels more slowly and constipation may result; it is more difficult to expel and this can lead to haemorrhoids, or piles; and without the satisfaction of a full stomach, provided by the fibre's bulk, a fibreless diet can lead people to eat more of the fattening foods and become overweight.

An important component of the large intestine is bacteria – a proportion of the body's massive population of resident micro-organisms there to do a job. In the large intestine bacteria manufacture some important vitamins, which are then absorbed into the body. By breaking down some of the fibre, they also contribute to the considerable volume of gases that lurk in the large intestine and occasionally leak out to the outside world. Of these gases, 99 per cent are odourless by-products of the bacteria's normal breakdown of fibre, but foods like onions and cabbage are resistant to our digestive enzymes and so they are not broken down until they reach the bacteria in the colon. It is probably during the extra breaking down of such foods that the small but intrusive fraction of odorous gases is added to the rest.

Once food is travelling through the large intestine, it is likely that the cycle has started all over again at the other end. It will be many hours and several meals later before the last components of breakfast see the light of day again – perhaps several days. In following the events of digestion stage by stage, it is easy to forget that, like so many bodily processes, they all take place simultaneously, with great efficiency and with the minimum of interference to our other everyday activities.

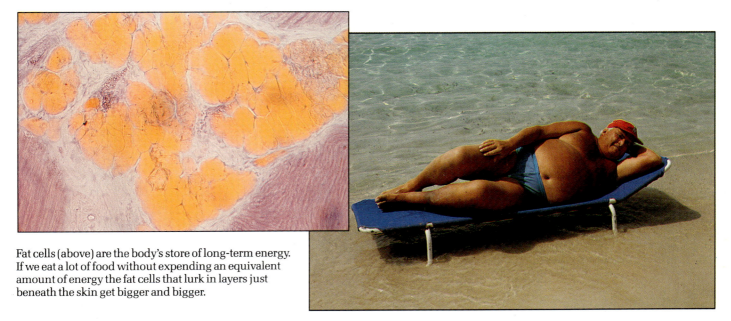

Fat cells (above) are the body's store of long-term energy. If we eat a lot of food without expending an equivalent amount of energy the fat cells that lurk in layers just beneath the skin get bigger and bigger.

6 · Water!

Thirst is one of the strongest drives the body possesses. If we are denied both food and water, we feel thirsty long before we feel hungry. Many people have survived 30 days' starvation without ill effects, but without water they could die within two or three days.

It is not really surprising that regular supplies of water are so important to us – after all, over half our body weight is made up of water, 40 litres (70 pints) of it, some of it in the blood and the rest contained in the cells and the space between them. Water is a transport system and a laboratory for many of the important chemical reactions that are needed to keep the body in good working order.

Because these activities are so important, the body has developed a system that controls the correct concentration of salts and the total volume of body fluids. Our output of water is roughly the same as our input. Between the two there is a delicate interplay of brain, kidneys, lungs and skin which makes sure that our bodies are neither too wet nor too dry.

We have evolved on a watery planet, and in our everyday lives we try never to be too far from water. During the events of a single day, the body can gain or lose water in a number of different ways – by sweating, breathing, eating, drinking, urinating and defaecating – and yet the internal watery environment of our cells stays remarkably stable.

If the body's water content drops, we soon start noticing the effects. After a 5 per cent loss, we become very thirsty; 10 per cent and we become very ill; and if the body's water resources drop 20 per cent we die, with 30 or more litres (over 50 pints) still left in the body.

The importance of salt

One important role for the water in the body is to carry various dissolved chemicals to the sites where they are needed. Salt (sodium chloride) is the most important of these chemicals. It plays a vital part in our physiological processes, and has an intimate relationship with the body's water supplies. If the balance between the salt

Excessive loss of water through sweating or evaporation in a hot climate can lead to illness or death. If the body's water level drops too far every cell in the body suffers as the blood becomes more and more concentrated with salt and other body chemicals. High concentrations of salt in the blood cause cells to collapse on themselves; conversely if the blood is too dilute cells expand. Red blood cells (inset) are particularly vulnerable; those on the left have collapsed and those on the right are dangerously swollen.

and the water is upset, it could mean disaster for the body's cells. The cells contain potassium chloride rather than sodium chloride, and the concentrations of potassium salt inside and sodium salt outside must remain at virtually a constant ratio. If this delicate balance changes, for example if the cell's surroundings become more dilute or more concentrated, then there is a pressure difference across the cell membrane, its outer covering. This pressure is called osmotic pressure and occurs in any situation where there are different concentrations of chemicals on either side of a semi-permeable membrane – a type of membrane that water can pass through but the salts dissolved in the water cannot. If the solution outside a cell becomes very dilute, water moves into the cell across the membrane and the cell will expand, sometimes bursting. Conversely, if the solution outside is more concentrated, water leaves the cell, leading it eventually to collapse. In other situations in the body, however, osmotic pressure can do useful work.

To prevent our cells exploding or imploding, we have developed a system that controls sodium concentration to within 0.5 per cent. This has to operate in the midst of human activities that can lead to the exchange of large volumes of water and salt between the body and the outside world. Three meals a day, for instance, could play havoc with the body's salt and water content. Whether or not we sprinkle salt or drink water with a meal, the food itself will have salt and water as components. And when, as well as dietary intake, there is the evaporation that takes place during breathing and the water loss of sweating, the ups and downs of water balance can be considerable.

Detecting the need for water

To monitor the changing situation, two types of detector cells in the body, osmoreceptors and pressure receptors, recognize when the blood becomes too concentrated or when the volume of blood decreases. When this happens it means that the body needs more water. The detector cells alert the higher centres of the brain, and we interpret the signals as feelings of thirst.

The concentration of salts in the blood can increase if we lose water or gain salt – both make us thirsty. The osmoreceptors in the brain come into action. But sometimes the body can be at risk from water loss even when the concentration does not change.

When we eat, the alimentary tract floods with digestive juices. These are watery and have been extracted from the blood, so they reduce the volume of blood in the circulation. This drop in volume also leads

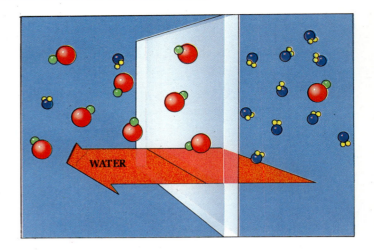

to feelings of thirst, which is probably why we often feel like a drink with our meals. Protein in particular needs a large volume of juices to digest it, so a large steak can soon make the eater thirsty.

A drop in blood volume can be very serious for the body – it may even lead to the heart being unable to generate enough force to circulate the blood. To provide an early warning of this danger, there are volume receptor cells in the heart and some blood vessels. These respond to stretching in the walls of the vessels or chambers. If we lose blood, in an accident for example, the detector cells will set in motion a series of events that conserve body fluids.

The drop in volume also leads to a strong feeling of thirst whenever we lose blood. This is a signal to us to replace the lost fluid as soon as possible. Because this fluid loss leads to water and salt being lost in equal amounts, it would not have been noticed by the detector cells in the brain. Water balance is so important that the body has evolved not one but two early warning systems working in harness.

Of course, we usually drink long before we are in any danger of dying of thirst. In fact, we normally drink more than we need, and the body excretes the excess. Drinking, like eating, has become a part of human social activities – we get pleasure from it. This is one respect in which man seems to differ from animals. They drink only when they need to. Both man and the animals seem to possess a 'water meter' in the gut which helps them to replace exactly the water they have lost, long before the other receptors have had a chance to register a change. However, man's social drinking habits override this mechanism. There is also an innate appetite for salt, to make sure we get enough. One of the four types of taste buds detects salt, and many animals automatically eat more salt if the level in the blood drops.

The simple diagram on the left illustrates the principle of osmosis: water molecules, being more concentrated on one side of a semi-permeable membrane than on the other, migrate through it to equalize the concentration. This is essentially what happens in the kidneys and large intestine, both specialized reabsorbers of precious water. In the kidneys, for example, there are special cells which concentrate salt in order to be able to reabsorb water and prevent it being lost in the urine. If the kidneys stopped reabsorbing water, which they do extremely efficiently, complete dehydration would occur within minutes.

We have to take in at least 1.7 litres (3 pints) of water a day to make good the water we lose from our lungs and skin, and in our faeces and urine. Only a third of this loss occurs through the kidneys; another third is lost through the skin, and about a sixth is lost by evaporation from the lungs. During profuse sweating an adult can lose up to 1.5 litres ($2\frac{2}{3}$ pints) of water an hour!

The endoscope photograph below shows a drink of water coursing through the gut. Digestion itself is a water-demanding process, since the mucus secreted by the oesophagus, stomach, duodenum and small intestine during digestion is largely water.

Disposal of water – the kidneys

Drinking deals with only one half of the body's water balance – it prevents us drying up from lack of water. But if we have *too much* water in our body then we must get rid of some of it.

This is not as easy as it might sound, because water plays an intimate part in all body processes. To extract water from the blood, the cells and the spaces in between is not a simple task. Once the water is in the body, it is mixed up with many useful chemicals, foodstuffs, hormones and so on. The body has had to find a way to dispose of surplus water without losing these more useful substances.

Extraction is the job of the kidneys, which act as a filter, but a discriminating filter, which can remove from the blood some of the water it contains, together with just those waste products we do not need. Although the kidneys act in some ways like a sieve, they can usually

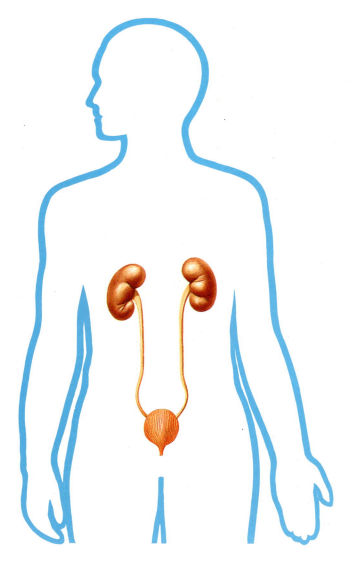

put back useful substances that slip through the holes.

We have two kidneys, although we could get by with one third as much tissue, and many people do. The kidneys are in a very good position to filter the blood, as they lie close to the main artery and vein that run down the middle of the body. With every heartbeat, the kidneys receive a quarter of the body's blood flow.

In the course of a day, the kidneys filter our total blood volume nearly 300 times; that amounts to 1000 litres (1750 pints) of blood. Out of all this blood, however, we produce about 1 litre ($1\frac{3}{4}$ pints) of urine a day, a very small proportion of the fluid that flows through the filter.

The work of the filter is really done at a microscopic level. More than a million tiny hairpin tubes are crammed together in each kidney, each dealing with a single droplet of blood. At the top of each tube is a basket of blood vessels called a glomerulus, which carries blood under pressure. The walls of these vessels do act like a sieve at this stage, passively letting through the water molecules in the blood, together with other molecules that are smaller than the holes. At this point, many of the body's soluble wastes, produced by metabolic processes, and toxic breakdown products of drugs, also pass through the glomerulus.

So the blood is split into two parts – water and chemicals that slip through the 'sieve', and the rest of the blood that stays in the capillaries. But these two fluids are never far away from each other. After the blood vessels leave the glomerulus they cling to the side of the hairpin tubules which carry the water and chemicals already extracted from the blood. The reason why these blood vessels and tubules are so closely linked is that some of the contents of the tubules is going to have to be put back into the blood, while the rest is eventually excreted as urine.

Glucose, for example, is a molecule that is small enough to slip through the glomerulus, and much of it would go down the tubule and into the urine unless something were done to stop it. What happens is that the walls of the tubule act rather like a pump and recapture the glucose, putting it back into the nearby blood vessels. This system is usually very efficient: more than 99 per cent of the glucose that tries to escape is put back into the blood.

But if we overload the system, as happens with diabetes for example, so much glucose leaks through that it cannot all be pumped back, and the surplus carries on down the tubule and out into the urine. Sugar in the urine is one way of detecting that we have the disease.

One of the other useful substances that might get

The kidneys lie close to the spine just in front of the root of the twelfth rib. They weigh about 140g (5oz) each and are buried in fat.

The kidney in the drawing on the right is about half life-size. Blood arrives through the renal artery (red) and flows towards the cortex or outer layer of the kidney. Here the blood vessels, now fine capillaries, bunch into little balls called glomeruli – lots of these are shown in the photograph below.

Hugging each glomerulus is a capsule that leads to a highly convoluted tubule, which then straightens, dives towards the medulla or middle of the kidney, makes a hairpin bend, and returns to the cortex. Here it convolutes again, then joins a collecting duct that meets up with thousands of others. These eventually emerge from the kidney as the ureter or urine duct.

At the bottom of the page is a corrosion cast of the blood vessels in a glomerulus (left) and also an artist's impression of a nephron, the glomerulus-capsule-convoluted tubule unit (right).

Below is a crystal of sodium chloride. The cells that line the capillaries and tubules in the medulla of the kidney concentrate salt in order to absorb water from the blood and urine.

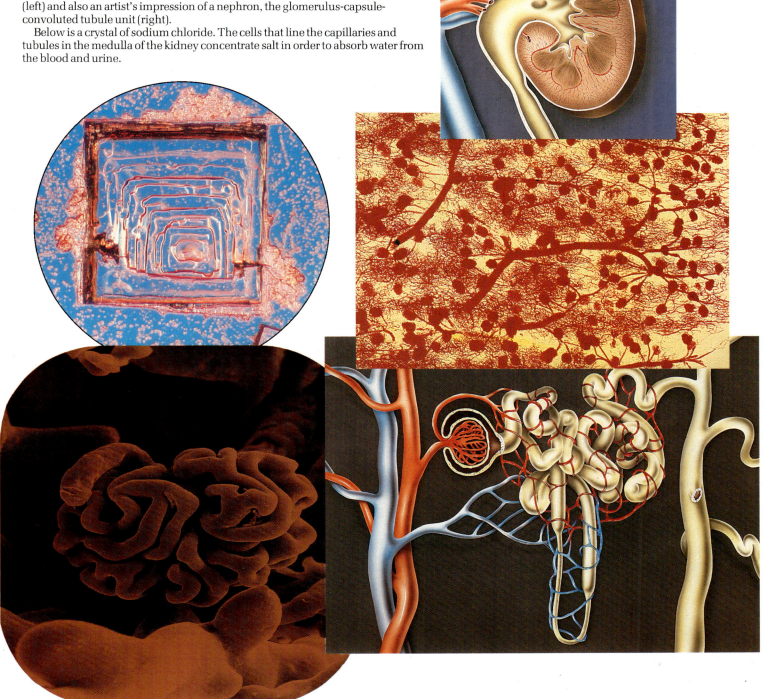

away is water itself. There would be little water left in the body if all the water that entered the tubule went on into the urine, since the equivalent of the body's entire blood volume passes through the kidneys about every five minutes.

This is where the intimate relationship between salt and water is very useful. Although salt is essential to the body, we can have too much of it. It is the job of the kidneys to correct our salt balance at the same time as dealing with the body's water supplies.

Salt, like glucose, is a small enough molecule to escape into the tubule. But we need to hold on to a sufficient amount of salt in the body and so, as with glucose, there are pumps in the walls to put the salt back into the blood. When this happens, the water in the tubule follows the salt back into the blood. This is because the salt that is pumped out of the tubule produces a concentration difference on the other side of the tubule wall, and so osmotic pressure drives the water through the walls to minimize the difference. It is effectively dragged with the salt, so that the concentrations either side of the tube wall stay roughly the same.

By the time the fluid that was filtered out of the blood has travelled all the way down the tubule, it has changed dramatically. Most of the water, salt and glucose has been put back into the blood. All that remains is a trickle of liquid – an amount that is just enough to correct any surplus of water we might have acquired by drinking – containing salt and water-soluble wastes.

So, from the individual hairpin tubules, the urine gathers in the centre of the kidney and flows into larger and larger collecting tubes.

The bladder

Up to this point we are unaware of any of this activity. The kidneys do not usually send us conscious messages and we have no control over their filtering. But there comes a time when we must make conscious decisions about fluid disposal: during the course of a day we need to urinate several times.

The urine first makes itself felt in the bladder, having entered in little spurts through the two ureters from the kidneys. These lead in at an angle and, as the bladder fills, they are squashed flatter. This helps to prevent

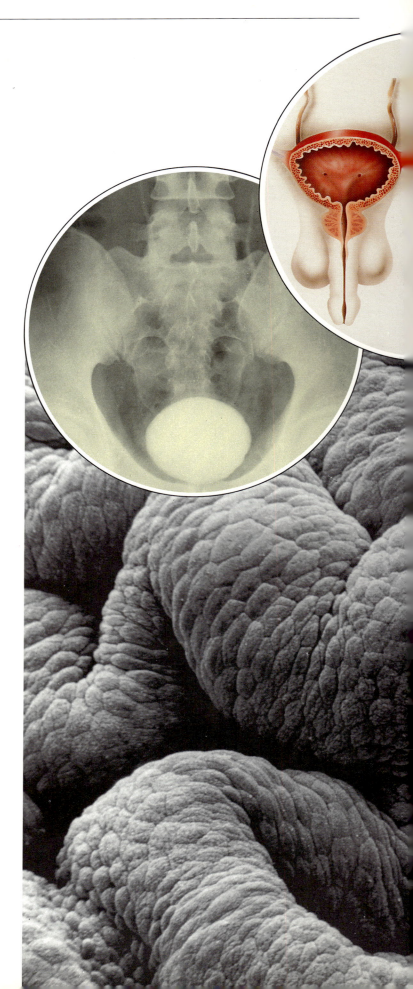

The fascinating undulations and folds in the photograph on the right are part of the inside wall of the bladder. A full bladder, shown in the X-ray above, holds about 0.6 litres (1 pint) of urine, but the urge to urinate begins when it is only a quarter full.

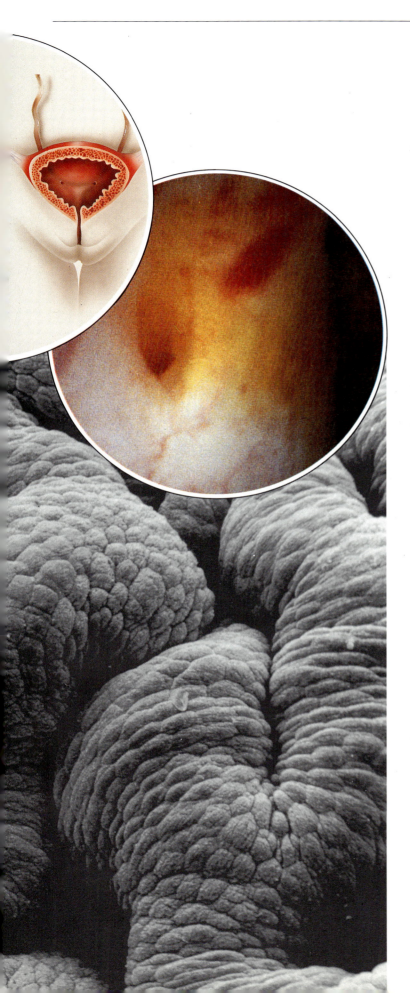

urine leaking back towards the kidneys.

We first feel a full bladder when it is only about a quarter full of urine. But if we do not act then, muscles and nerves in the bladder wall stop sending signals for a while. As more urine trickles in, the bladder again sends signals of discomfort but, again, we can choose to do nothing about it. In this way, we can choose the moment of urination, rather than having to empty the bladder automatically.

Sometimes we can last quite a long time before we feel the need to urinate. At other times, the bladder seems to fill much more quickly. It all depends on the other means we have of decreasing the body's water content. On a hot day, for example, we lose a lot of water by sweating, so the kidneys do not need to filter so much water out of the blood. That is why we need to urinate more often when it is cold – urination then makes a larger contribution to water loss.

As the messages from the bladder become more insistent we choose a time and a place to urinate. The exit from the bladder has two successive bands of muscle. One opens automatically when the bladder reaches a certain pressure. But beyond the first band is another muscular ring, under voluntary control. As we mature we learn to keep this closed until the appropriate time. Then when the conditions are right we relax the muscle and urinate.

The final stages of urination are familiar to all of us. But there are minor differences depending on our sex. In men, the urethra, the tube that leads from the bladder to the outside via the penis, is up to 20 centimetres (8in) long. In women, there are only 3-4 centimetres ($1\frac{1}{4}$-$1\frac{1}{2}$in) between the bladder and the outside world. This is thought to be why women get bladder infections more easily than men – the bacteria have a shorter journey from the outside to the bladder.

For all of us, water input must balance water output. However much we drink in a day we have to dispose of in a day, somehow or other. By eating and drinking we look after the input; by urination, defaecation, sweating and breathing out we carefully adjust the output.

The purpose of all this activity is to allow the body's cells to get on with their everyday activities, without being poisoned, drowned or dehydrated, in a calm and peaceful internal environment.

At the top of the page are the male and female bladders, showing the relative lengths of the urethrae and the points of entry of the ureters from the kidneys. The prostate gland beneath the man's bladder adds fluid to semen during ejaculation.

The endoscope picture to the right shows urine entering the bladder.

7 · Growing Concern

Some animals grow as long as they live and live as long as they grow. But for the human animal, growth stops when we are less than one quarter of the way through our life. During this time, we progress from a helpless four-kilogram baby to a mature and self-sufficient adult.

The most obvious sign of this dramatic expansion is height. We measure children year by year and stop doing it when it seems that they are not going to grow any more. Both of these phenomena – growing and stopping growing – raise various questions, not all of which can be answered.

During the course of our lifetime, we must grow from a fertilized egg to an adult of the correct size and maturity to survive and reproduce in the world on our own. Growth in the womb is important to achieve the first stages of development. But in spite of the tremendous speed and complexity of growth, the new-born baby is only a twenty-fifth of the weight that it will eventually grow to – for the next 15 or 20 years every part of its body must enlarge and change shape in an orderly manner. And this task involves a carefully programmed sequence of events that constitute far more than an increase in size. There are huge differences in proportion between a baby and an adult. Head size, for example, is one quarter of the body in a baby and one eighth in an adult.

This means that the processes of growing up must control shape as well as size – hands grow in a carefully programmed way, rather than just expanding outwards in all directions. And the complexity of the human face shows the body's ability to preserve shapes and proportions throughout a whole lifetime: a face can be recognizably the same for 70 years. Presumably, it has something to do with the relative position of the eyes, the nose and the mouth, and perhaps the shape of the head, but whatever it is, when we have become familiar with it in an adult, we can then recognize it in photographs of the child he or she once was.

There is in fact little physical continuity between the body of an adult and his body when he was a baby. Most of his cells have died and been replaced many times over. How, then, does his body preserve the shapes and sizes of his organs over a lifetime, as they grow in three dimensions?

How do we grow?

Growth depends on individual cells. Every organ and body system is made up of cells or the products of cells, and a full-grown human is made up of about 30 million million cells. Since we all started from only one cell, the fertilized egg, a very great deal of multiplication has obviously taken place.

An increase in numbers of cells comes about by cell division. Each cell splits into two usually identical cells. But, although nearly every cell in the body contains identical genes handed on from the first cell (the fertilized egg), each cell may develop into a very different part of the body. This is possible because every cell contains in its genes a blueprint for building the entire human body. While two cells in the earlier stages

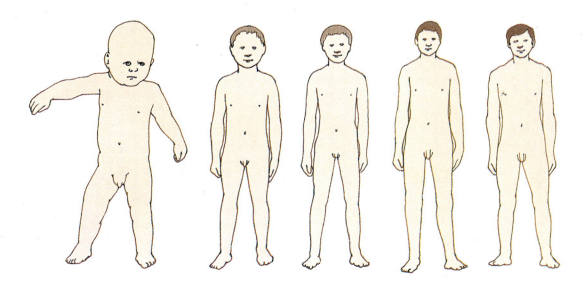

At birth, a baby is generally between 27 and 30 per cent as tall as the adult he or she will become. In evolutionary terms this represents a compromise between the largest size that can be sustained advantageously in the womb and the smallest size that is viable enough to have an excellent chance of survival. By the age of two a baby will have reached about half final adult stature; by five about two-thirds. Girls commonly achieve their final height at 15, boys more often at 18. The drawings on the left show the relative proportions of head and limbs to body size from babyhood to adulthood.

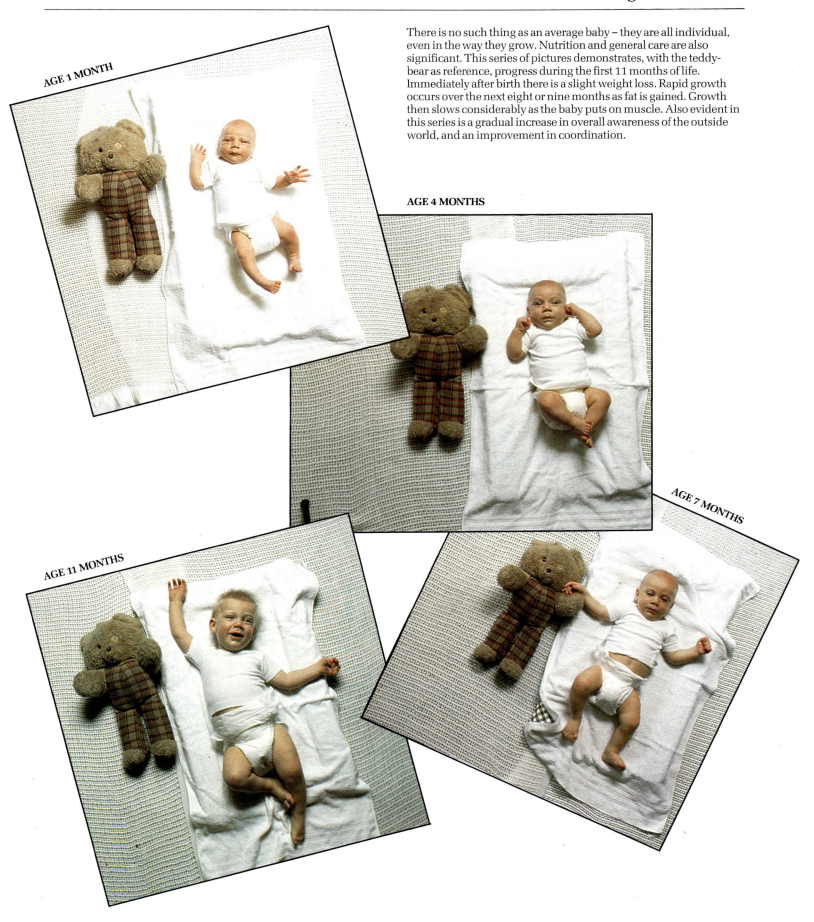

AGE 1 MONTH

AGE 4 MONTHS

AGE 7 MONTHS

AGE 11 MONTHS

There is no such thing as an average baby – they are all individual, even in the way they grow. Nutrition and general care are also significant. This series of pictures demonstrates, with the teddy-bear as reference, progress during the first 11 months of life. Immediately after birth there is a slight weight loss. Rapid growth occurs over the next eight or nine months as fat is gained. Growth then slows considerably as the baby puts on muscle. Also evident in this series is a gradual increase in overall awareness of the outside world, and an improvement in coordination.

These pictures of the much-loved Queen Mother show how constant the main features of the face are despite the passage of time. Much of that constancy is dependent on skin tone. The skin is a complex, layered formation of cells, waterproof yet full of pores; it regulates body temperature and is very adaptable, sensitive yet tough. Support for all the tissues of the face is provided by collagen, which becomes more elastic as age increases. A looser skin loses its rounded contours and begins to wrinkle.

Although there are many types of cell, all are basically made up of a nucleus surrounded by cytoplasm enclosed in an outer membrane, as seen at the top of the diagram opposite. The cytoplasm contains nutrients, basic building materials, wastes and various tiny organs or organelles. The cell nucleus contains all the information necessary for the cell to grow, perform a specific function, or duplicate itself. This information is coded into discrete packages called chromosomes, each containing hundreds of specific instructions called genes.

of the embryo are identical, they contain all the information necessary to become *any* cell. The future paths of these cells will diverge as, for instance, one becomes a nerve cell and the other becomes a bone cell. They seem to do this by each switching off the irrelevant part of the blueprint and leaving active only the relevant genes to control their future development towards nerve or bone.

This combination of the process of cell division with each cell's potential to become any type of cell produces growing tissues in the human body – tissues which are specialized for particular tasks like developing power, passing on messages or synthesizing hormones. So a small blob of identical cells changes over time into a beautifully organized, fully functional limb or organ.

But while some organs and body tissues grow by multiplication of cells, others grow by individual increases in cell size. In muscle tissue, for example, virtually the whole increase in mass from infancy to adulthood is a result of the same number of cells increasing many times in size.

As we only ever see individual cells under a microscope, it is difficult to appreciate their actual sizes or the extreme variations in size. The largest cells in the body, the nerve cells that lead from the spinal cord to some of the muscles, can be up to one metre (3¼ft) long, while a sperm is only 1/2500 of a millimetre.

So growth is really the combined result of cells increasing in number *and* size. But that is just a starting point. These increases must be coordinated to produce the correctly shaped end result, in the correct part of the body. Something like an arm, for example, must start growing in the right place – out of the shoulder rather than the hip or neck.

Development of the limbs and organs

Both inside the womb and outside it, our limbs and organs grow in an orderly and shapely way. Throughout our adult lives, for example, our arms stay roughly the same length as each other. This may not seem very surprising at first, but there is no obvious reason why they should: the growing cells of skin, bone and muscle

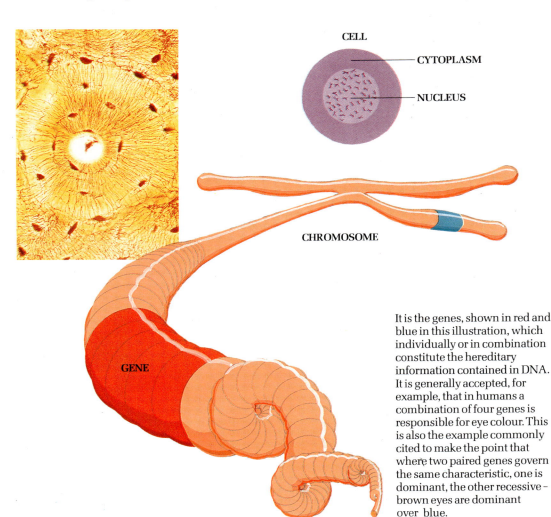

CELL

— CYTOPLASM

— NUCLEUS

CHROMOSOME

GENE

It is the genes, shown in red and blue in this illustration, which individually or in combination constitute the hereditary information contained in DNA. It is generally accepted, for example, that in humans a combination of four genes is responsible for eye colour. This is also the example commonly cited to make the point that where two paired genes govern the same characteristic, one is dominant, the other recessive – brown eyes are dominant over blue.

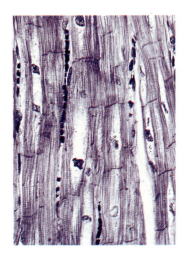

But more specifically, it is the genes which signal to the cell around them what type of cell it is to be, how it is to grow, whether it is to replicate, and what function it is to perform. For example, bone cells, shown on the left, have long spiky 'arms' filled with cytoplasm that mould minerals into the strong, rigid struts that form our skeleton; muscle cells, like the heart muscle cells shown above, have contractile filaments in them.

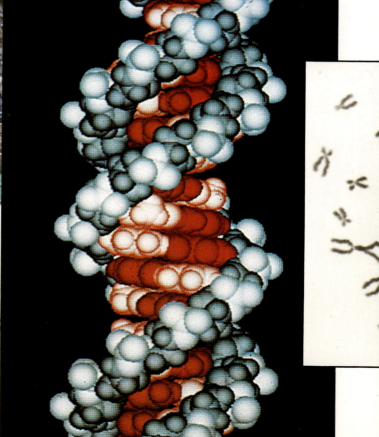

The colourful pattern above was part of the opening ceremony at the 1980 Olympic Games in Moscow; acting on instructions, the card-carrying crowd varied the patterns. In a similar way our genes can be instructed to display different characteristics.

The chromosomes in each cell contain a complete 'plan' of the *entire* body, yet each cell has only one function and only one location. It is not yet fully understood how each cell fulfils its specific role, and ignores all the other elements of the plan. But however the system works, it works extremely well almost all of the time, and is especially remarkable at the edges of defined areas – for example, outer layer bone cells are 'one-sided' compared with inner layer bone cells.

in each arm are widely separated from each other and have no obvious links to coordinate their speed of growth so accurately. And yet, somehow, at each stage in the development of the human, each cell knows what to do, where and when to do it, and when to stop.

It is not known for certain how these cells are controlled in such an organized way but there is a plausible theory for how it might happen.

In a developing limb, each cell needs one piece of information in addition to its collection of chromosomes. It needs to know which genes to switch on while the others are left dormant. That event, the switching on or off of genes, is likely to depend on the cell's position in the limb. After all, a cell's role in the limb, whether it is bone, skin or fingernail, is usually related to its position.

It is then possible to think of a system by which some factor, perhaps the varying level of a chemical, indicated where a cell was, and that in turn determined which parts of the chromosomes should be activated.

There is some evidence in animals that development happens in this way, and it is even possible to 'deceive' cells in a growing limb that they are in a different position. When that is done, the cells then take on a new identity and play the role that the cells in its new position are supposed to play. It is as if we were able to persuade a fingernail cell that it was only in the position of the knuckle. If the trick worked, and it does in some amphibians and lizards and in some other groups of animals, then it would start to grow the bones and skin of a finger instead of finishing itself off as a fingernail.

It is sometimes the case that things that go wrong with the body can lead us to a much clearer understanding of how it works normally. With human growth, faulty development can lead to people growing too tall or ending up too short. With short stature, there is more than one type of error, and two people of the same restricted height can show important differences. One may be half the average human height but with normal proportions. Another, although the same height, might have a different set of proportions. His head, for example, could be the same size as a normal adult head and the girth of his limbs and torso be the same as those of a man of normal stature. The reasons for these differences become clearer when we look at how the skeleton is formed and grows. For this is the most important factor in determining body size.

In a very few people, the genes that code for the production of certain hormones may be faulty. Too much or too little growth hormone produced at the critical time leads to giantism or dwarfism. The photograph on the left shows the two extremes.

On the left is a computer graphic model of DNA, showing its double helix structure. The blue molecules are sugar and phosphate.

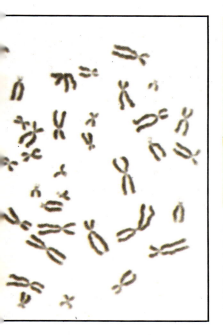

Count the chromosomes shown on the left and you will find 46, the full complement of *Homo sapiens*. But occasionally individuals are born with 45, or 47 or even 48.

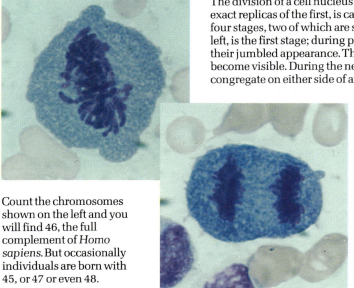

The division of a cell nucleus to form two separate cells, both exact replicas of the first, is called *mitosis* and takes place in four stages, two of which are shown here. Prophase, on the left, is the first stage; during prophase the chromosomes lose their jumbled appearance. Then individual chromosomes become visible. During the next stage, metaphase, they congregate on either side of an 'equator'. Then comes anaphase, and the chromosomes begin to draw away from the equator – this is the stage shown on the left. In the final stage, telophase, the cell splits across the equator. Result: two daughter cells identical to the original cell.

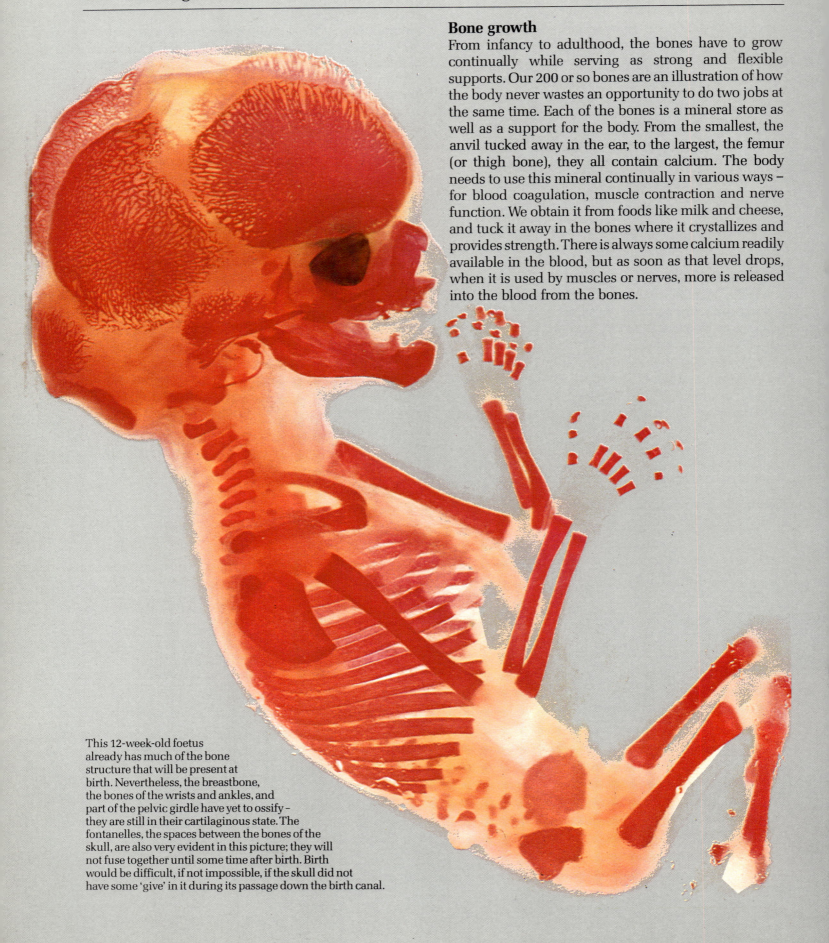

Bone growth

From infancy to adulthood, the bones have to grow continually while serving as strong and flexible supports. Our 200 or so bones are an illustration of how the body never wastes an opportunity to do two jobs at the same time. Each of the bones is a mineral store as well as a support for the body. From the smallest, the anvil tucked away in the ear, to the largest, the femur (or thigh bone), they all contain calcium. The body needs to use this mineral continually in various ways – for blood coagulation, muscle contraction and nerve function. We obtain it from foods like milk and cheese, and tuck it away in the bones where it crystallizes and provides strength. There is always some calcium readily available in the blood, but as soon as that level drops, when it is used by muscles or nerves, more is released into the blood from the bones.

This 12-week-old foetus already has much of the bone structure that will be present at birth. Nevertheless, the breastbone, the bones of the wrists and ankles, and part of the pelvic girdle have yet to ossify – they are still in their cartilaginous state. The fontanelles, the spaces between the bones of the skull, are also very evident in this picture; they will not fuse together until some time after birth. Birth would be difficult, if not impossible, if the skull did not have some 'give' in it during its passage down the birth canal.

We can see how this works if we take one long bone like the thighbone. This increases in length by cell division and by increase in cell size. In fact, the bone substance itself is made up of flexible fibres combined with minerals laid down by bone-forming cells. It is really a non-living crystalline framework inhabited by living cells. This combination of flexibility and rigidity makes bones very strong and resilient.

The landscape of bones is a series of cylindrical potholes, down which run vessels to carry blood and lymph. Around each cylindrical hole are rings of mineral salts like walls of rock, riddled with holes in which lurk the bone cells. One group of these, the osteoblasts, is responsible for building bone, by making fibres of collagen (a connective tissue) and depositing calcium on them. The other group are called osteoclasts and they are able to break down collagen and release calcium into the blood stream. With these two groups of cell hard at work in our bones, there is constant activity as one destroys the work of the other. It may seem to involve a lot of unnecessary work, but in fact it makes a lot of sense.

First, by slightly altering the balance of activity between the two groups, the body can control very accurately the amount of calcium that is available for various bodily processes. If more calcium is needed, the destroyers start liberating calcium from bone at a faster rate than the builders absorb it.

Secondly, our bones are not meant to be static and unchanging. Even after we have grown to our full size, there is a need for our bones to change continually in response to the physical demands placed on them. The more a particular bone is stressed, the more active the bone-building cells will be, to strengthen the bone at the

The newt above is in the process of regrowing a hind limb. We humans have lost this convenient capacity, although we regenerate individual cells in tissues like bone, liver and skin on a daily basis.

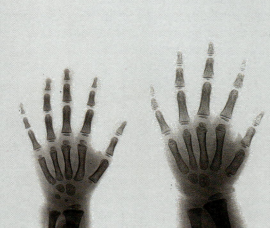
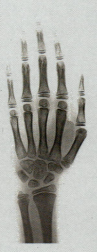
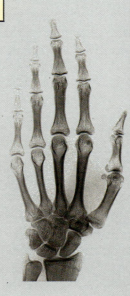

These four pictures show the slow development of the wrist bones in a baby, a toddler, an adolescent, and an adult. Only in the adult are the carpals truly all bone.

point of stress and to repair microscopic breakages that are occurring all the time as a result of physical effort.

While the body is developing, it is always the builders who are working harder than the destroyers, lengthening and shaping the individual bones. The direction of the growth is guided by a thin skin that surrounds the bone as it develops. Rather like a corset, this skin confines the developing bone cells so that it is much more difficult for them to expand outwards than to grow lengthways.

Soon after birth, the ends of the bones become solidified and an area in the middle also loses the ability to grow in length. From then on, long bones grow in two areas near the ends of the bone, pushing the ends further away from the middle. A soft framework of cartilage joins the hard end of the long bone to the bony middle section. This is called a growth plate, and as it lengthens, the face turns into hard bone.

So long as there is still some growth plate left, the bone can continue growing. It is only when those areas have shrunk to nothing, at about 20-24 years of age, that the bones stop growing.

Sometimes, bone growth is much faster than it should be and the bones reach a much greater length before the growth plates shrink away. This is how giants are created – not because they go on growing after others have stopped but because they grow faster over the same period.

So there are two factors that are important for setting the eventual length of a bone – the rate of growth, and the time at which the growth plates disappear. Two short people may suffer from different types of defect in bone growth. One, with normal proportions, may have bones that grew at a slower than normal rate. Since his growth plates closed at the normal time, he remained shorter than average. The second may have a condition that caused his growth plates to close earlier than normal. His ability to grow bone at the normal rate was unimpaired but there was no growth plate left to expand, at least in a lengthways direction. Because his bone was still able to grow outwards, by a different mechanism, he acquired the dimensions of an adult but not the height.

What determines our adult size?

Although the growth of all creatures is determined by the rate of growth and the time for which it continues, the two interact in different ways. Rabbits are bigger than guinea-pigs, for example, because they grow faster, but humans are bigger than rabbits because they grow for a longer time. In each species these characteristics are controlled by hormones, produced by the brain and the sex organs. There is a growth hormone (from the brain) which controls the rate of growth of bones and other organs, and there are other hormones, called androgens (from the sex organs) that close the bone growth plates. In humans, androgens are responsible for many of the bodily changes that happen at puberty and help to set the final size of the adult by locking each of the long bones at a certain size.

While the bones are growing, other organs are trying to keep pace. Muscles, connected to bone, adjust their sizes according to the bone length, and the major organs in the body obviously have to keep pace with the increase in body size. Sooner or later, however, all this well-coordinated activity comes to a halt, as we reach our adult size.

The ultimate size that each of us grows to depends largely on our parents. In humans there are well-known differences in size that are obviously inherited and show themselves in families or races; our genes contain a lot of information about the rate at which we should grow and the point at which we should stop. Fortunately, the information about our ultimate size does not start influencing our growth too early. There is not a lot of correlation, for example, between the size of a baby and its eventual height (although by childhood there may be). In fact it would be very inconvenient if there were: tall men sometimes marry small women, and since the father's genes controlling size could well dominate in the child, it might be too large to leave its mother's womb. Fortunately, the size of a child at birth seems to have much more to do with the size of the mother than the father. It is possible, for example, for a large Shire stallion to mate with a tiny Shetland mare, and produce offspring which fit easily into the tiny uterus of the Shetland even if the embryo is eventually destined to grow to several times the size of its mother.

Man's growth differs from animal growth, however, in the length of time the human offspring take to reach adult size. Compared with many animals, we go on growing for far longer after we have left the womb. It might seem a disadvantage for our species to have such a long childhood and be dependent on our parents for so long. But it is actually an advantage: the young human has much more time to develop intelligence and acquire skills, by imitation of his elders and by a fertile use of his rapidly developing brain. The characteristics of children that we all recognize, and sometimes find infuriating, are actually an important part of becoming adult. As well as a readiness to learn, they include playfulness and curiosity – very important for adapting to a changing world. And some adults, even when they have grown to full size, retain these valuable characteristics in their everyday lives.

Adult stature is
mainly achieved by
growth of the long bones
of the body. Near each end of a
long bone, separating the top and bottom
from the middle section, is a growth plate where new
bone is formed. The ends of the bone recede from each other
until the growth plates disappear between the age of 20 and 24.

The knee joint shown at the top of this page is that
of a six-month-old foetus; the ends of the bones in
the lower leg and thigh are still cartilage, and so is
the knee cap. Despite the presence of so much
cartilage and connective tissue in place of true
bone in young people, the skeletal structure is
strong. Bone has an inner honeycomb structure
and a compact outer structure – minimum weight
with maximum strength. The photograph above
shows the concentric ring structure in the
compact outer region of a limb bone. Here bone
cells arrange themselves in rings around central
canals which carry blood vessels. The matrix
secreted by the bone cells gradually hardens as
crystals of calcium phosphate are deposited in it.

8 · It's Only Your Nerves

The link between a simple reflex action – such as the sudden withdrawal of a foot or a hand, or standing up straight without toppling over – and the cultural and intellectual achievements that set man apart from other animals is not obvious. And yet all our achievements are possible because of the way nerve cells behave, enabling us to respond promptly, and without our conscious control, to events in the world around us. Every part of the body is infiltrated by nerve cells, which form a communications network throughout the body. The most important nerves flow to and from the brain. Some carry sense messages; others carry orders from the brain. There is a major trunk of nerve pathways running

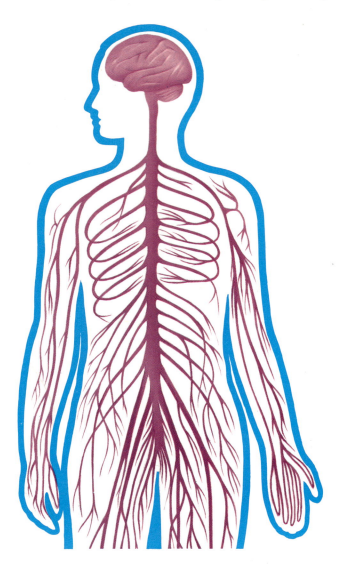

up the spine, the spinal cord, and nerves from the outlying parts of the body meet up there and travel together to the brain. Beside the spinal cord are small knots of nerve cell junctions which sometimes act in conjunction with it, like subsidiary brains, helping to make decisions that do not need to be referred up to consciousness. Every bodily activity begins as a message in a nerve cell, and the way those messages are passed on from one cell to thousands of others creates the special characteristics of the nervous system. By understanding how as few as *two* of those connected cells work, in a simple reflex, we take the first step towards understanding how our 100 000 million of them make human achievement possible.

From the moment we are born, we behave. Brightness and loud noises startle us; the voice and face of our mother soothe us; hunger agitates us and food calms us down. In each of these situations, we perceive something in the outside world and act on it – that is what behaviour is. Like everything else the body does, we are increasing our chances of *survival* by the behaviour we choose. The young baby seeks survival in its most basic sense – food and attention are the guiding principles. As he gets older, his behaviour may not appear quite so rational, but sooner or later what we do in this world can be traced back to our need to survive in it, either as individuals or as a race.

Yet behaviour depends on a body system – the nervous system – whose workings are not easy to observe. While we can see skin, feel bone and muscle, and even sense our heart beating, we cannot discover much about our brain, spinal cord and nerves in a similar way. Although this communications network of millions of cells is what makes us aware of everything else, we are not aware of *it*.

Intimately connected to every part of the body, this decision-making unit receives information about the outside world and decides what to do about it. It is incapable of acting on its own – a nervous system without a body would be as useful as a radio without a loudspeaker. But once it is linked to sensitive sense organs on one side and versatile muscles on the other, it makes the human being the most talented creature in the world.

Of course the most important part of this system is the brain – it supplies the creativity, the consciousness and

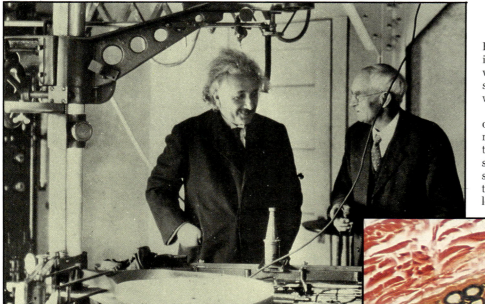

Each nerve cell has a body with a nucleus inside it and one or more long extensions called axons, which join up with other nerve cells. A nerve is simply lots of nerve cells running side by side with their axons parallel.

The photographs below show what bundles of nerve cells look like under high magnification. Immediately below is a slice through lots of axons and their insulating sheaths. At the bottom of the page the slicing shows a number of nerve cell bodies and also the parallel wavy patterns of axons running longitudinally.

The human brain consists of thousands of millions of nerve cells capable of an infinite variety of intercommunications. Without this variety there would be neither individual human intelligence nor creativity – neither the brilliant theories of Einstein, seen above at the Mt Wilson Observatory in the company of R.R. St John, nor the intricate fugues of J.S. Bach.

On the right is a cross section through the spine, vertebra and all. The spinal cord sits in the middle of the vertebra, protected by three membranes and by a special fluid secreted in the brain. The white area in the centre of the picture is the vertebral canal; the two-lobed structure in the middle is the spinal cord. Because this is a section from the lower spine the spinal cord can be seen surrounded by various nerves entering and leaving it.

The human nervous system, outlined in the drawing on the left, comprises the central nervous system – the brain and the spinal cord – and the peripheral nervous system – 31 pairs of nerves which branch off the spinal cord and link it and the brain with the rest of the body. There are also 12 pairs of cranial nerves which arise directly from the brain and relay messages to and from the eyes, ears, nose, mouth, face and so on.

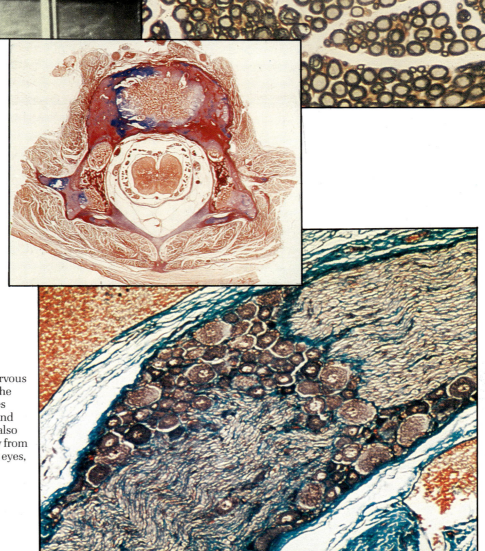

the intelligence of the human being. But there is a part of the nervous system that can do a great deal on its own without any need to refer to the brain: this is the spinal cord, which hangs below the brain, protected in the bony tube of the backbone.

Man and the other mammals have become so highly developed that they have acquired two levels of behaviour: situations which are unpredictable are handled by the brain and usually require some voluntary action; those which are either more mechanical or more urgent are left to simpler reflex systems involving the spinal cord, and sometimes the brain, and we are often not aware of them. Human reflexes need some outside stimulus to trigger a predictable response. The starting point could be some foreign body in the nose or the airways – passages that must be kept clear. Or it could be something that threatens injury. The trigger could also be some change in the environment that we need to respond to – brighter light means we have to constrict the pupils, or a change in body position means we have to adjust a few muscles to keep ourselves upright. In all of these cases some message that has come *into* the body requires a response *from* the body. And the response is predictable enough for simpler parts of the nervous system to deal with it, although the brain may also be notified of the event if skin receptors are involved – for example, although we withdraw our hand from a sharp object by reflex, the brain will also register pain.

How a simple reflex works

The well-known knee-jerk, or stretch reflex, is the simplest example of reflex activity, involving just two sets of nerve cells. Although we do not go through life jerking our knees to the tap of a hammer, this reflex plays a role in a very important activity – standing up. A tap on the knee has a similar effect to a sudden pull on the muscle along the top of the thigh, because it triggers a nerve message from stretch receptors at the end of the muscle. Similar nerve messages often occur in the thigh muscle and others when we are standing up. To stand up straight without falling over is actually more difficult than it seems; we are essentially unstable individuals, with a narrow base and a lot of weight that has to be poised vertically above it. And yet we can maintain an upright posture while doing all sorts of things that should upset the balance: waving, carrying, bending over – all of these would cause us to topple over unless our leg muscles were making constant adjustments in tension and position.

This control of posture is based on the stretch reflex, but involves more muscles. If we look at what happens

in one of the thigh muscles, we can begin to understand the process. One long nerve cell joins a stretch receptor organ in the extensor muscle at the front of the thigh (which ends in a knee tendon) to a part of the spinal cord. If something happens that might send the body off balance, the slight shift in position stretches the muscle and triggers a message from the receptor, along the nerve cell, to the other end in the spinal cord. This signal jumps across a small gap to another long nerve cell which carries the message back to the muscle, causing it to contract to adjust for the change in posture. At the same time, impulses to the muscle in the back of the thigh, the flexor muscle, are inhibited in the spinal cord to allow the extensor to contract. Of course this is not a one-off incident – all muscles involved in posture are sending and receiving impulses from several stretch receptors, and their antagonist muscles, which work in opposition, are doing the same.

Because the knee-jerk reflex uses important parts of the nervous system, it can serve as a valuable indicator that everything is functioning properly in the spinal cord, which is why doctors stimulate it artificially when trying to diagnose what is wrong with a patient.

A nerve junction is known as a synapse. At synapses in the spinal cord, as in the rest of the body, the first nerve cell ends very close to the second, but not actually in contact with it. As the impulses arrive at this point, they have to trigger the second cell to fire: for the knee-jerk reflex to work, the message has to be carried across the gap between the two cells. Although one electrical message triggers another electrical message in the neighbouring cell, it does not usually do it electrically. What happens is that messenger molecules are released by the end of one nerve, travel across the gap to the beginning of the other nerve, and trigger an electrical impulse that moves away down the nerve fibre. The way this takes place for these two cells applies to many nerve junctions in the body although different transmitter chemicals are used. There are also some junctions across which electrical transmission does occur.

Communication in the nervous system

The way in which a signal is transmitted along one nerve cell, and how it jumps from one to another, does not vary much, whether we are just trying to stand up straight or performing higher mathematics. The method of communication used by the nervous system is much simpler than any human language.

Man in society has managed to achieve communication over a distance in a variety of ways. How he does it depends on the complexity of the message, the distance to be covered and the likelihood of interference.

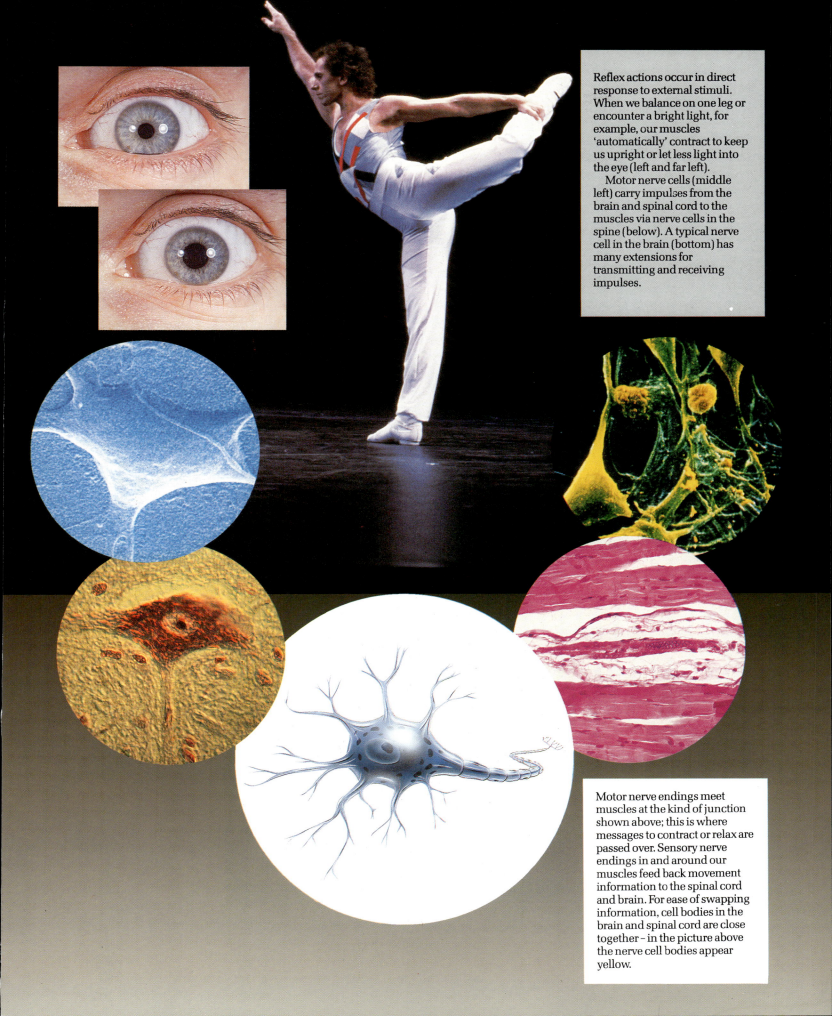

Reflex actions occur in direct response to external stimuli. When we balance on one leg or encounter a bright light, for example, our muscles 'automatically' contract to keep us upright or let less light into the eye (left and far left).

Motor nerve cells (middle left) carry impulses from the brain and spinal cord to the muscles via nerve cells in the spine (below). A typical nerve cell in the brain (bottom) has many extensions for transmitting and receiving impulses.

Motor nerve endings meet muscles at the kind of junction shown above; this is where messages to contract or relax are passed over. Sensory nerve endings in and around our muscles feed back movement information to the spinal cord and brain. For ease of swapping information, cell bodies in the brain and spinal cord are close together – in the picture above the nerve cell bodies appear yellow.

While human language is very good at conveying the subtleties and nuances of human life, it has certain drawbacks in situations where there is interference or 'noise' that might distort the message. There is a well-known party game where sentences are passed on in whispers down a line of children. In this way, 'send reinforcements, we're going to advance' gets transformed into 'send three and fourpence, we're going to a dance'. This sort of misinterpretation happens because there is a wide range of possible messages for the listener to choose from. If the people in the chain knew that the message could only be either 'apple' or 'pear', the correct message would probably travel from one end of the chain to the other without distortion.

There are many situations where a simple code makes it possible to get messages through against a background of noise. A tick-tack man on a racecourse wants to be understood over a great distance and so he uses a system of signals that leaves no room for ambiguity. He sends messages by a code that uses a small number of arm positions. It may take longer to send a message but there will be no doubt about what signal he is sending – one signal cannot be mistaken for another. With such a limited range of signals, it would not be advisable to choose to use this system to transmit *Hamlet*, but it could be done if there was time. With more arm positions the tick-tack man could transmit his messages in less time, but there would be much more opportunity for confusion, and bad light and human variation could add to the misunderstanding.

In an environment like the body, alive with electrical and chemical activity all the time, the language that nerves use *has* to be simple to be interpreted unambiguously through considerable interference.

To understand the language of nerves, we need to observe just one nerve in action. When we do that, we find a rather uniform series of messages. There are no specific instructions to 'contract the muscle', 'turn off the tap', 'shriek' or whatever is appropriate, but a succession of similar waves of voltage change passing down the fibre. These waves can be heard as clicks if the nerve is connected to a loudspeaker and the frequency with which they occur usually depends on the intensity of the stimulus.

An individual nerve cell will not transmit any message if the stimulus – heat or pain, for example – is

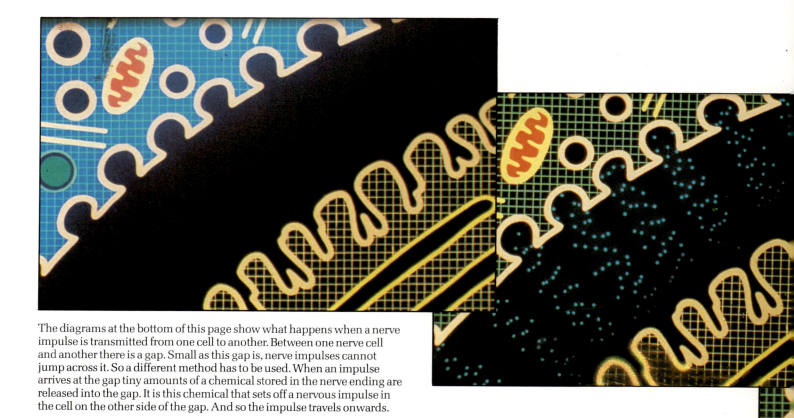

The diagrams at the bottom of this page show what happens when a nerve impulse is transmitted from one cell to another. Between one nerve cell and another there is a gap. Small as this gap is, nerve impulses cannot jump across it. So a different method has to be used. When an impulse arrives at the gap tiny amounts of a chemical stored in the nerve ending are released into the gap. It is this chemical that sets off a nervous impulse in the cell on the other side of the gap. And so the impulse travels onwards.

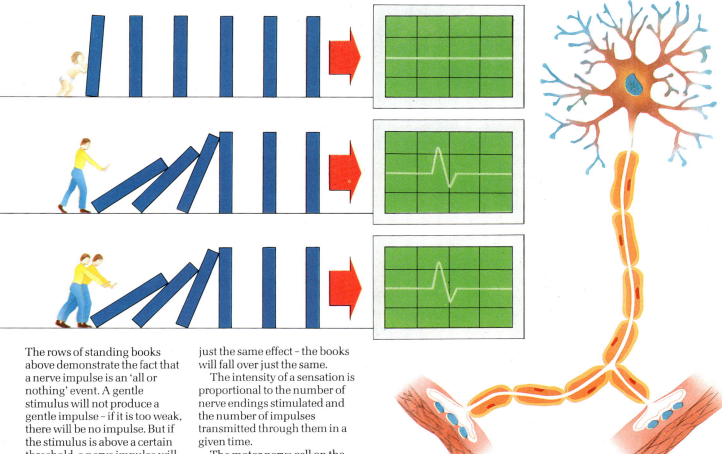

The rows of standing books above demonstrate the fact that a nerve impulse is an 'all or nothing' event. A gentle stimulus will not produce a gentle impulse – if it is too weak, there will be no impulse. But if the stimulus is above a certain threshold, a nerve impulse will be generated – the books will fall over. But make the stimulus twice as strong, and it will have just the same effect – the books will fall over just the same.

The intensity of a sensation is proportional to the number of nerve endings stimulated and the number of impulses transmitted through them in a given time.

The motor nerve cell on the right has many endings and a myelin-coated axon for carrying fast messages to muscles.

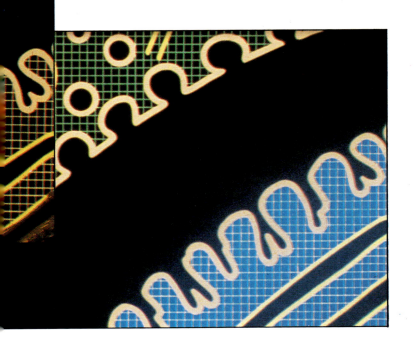

below a certain level. If the stimulus increases in strength then, when it reaches the necessary level, the nerve suddenly 'fires', sending out an impulse that causes the next cell to fire, and so on, like a chain of dominoes falling. As with dominoes, if the initial push is enough to start things off, the message will be as strong at the end as it was at the beginning. An individual nerve can carry a whole train of impulses, at a frequency of up to about 1000 times a second, as if the dominoes could be set up and knocked down again at great speed. So, although the strength of the initial impulses cannot be varied, the *rate* of firing can alter, and it is the frequency of the impulses rather than their strength that determines the intensity of the message. With light arriving at the eye, for example, the brighter the light, the more impulses per second travel along the nerve from the eye to the brain. Because it is only the rate that matters, this sort of message, unlike the whispered sentence in the party game, is not easily distorted on its long journey from one end of the body to the other. The *intensity* of each impulse can change due to distortion or interference on the way, but this will not affect the information, which will still be the same when it is interpreted at the other end. The only way a message can be interfered with is by damaging the nerve, which prevents any message being conveyed.

Speed of nerve messages

We are used to thinking of nervous activity as rather fast. Something like the reaction to pain or to a sudden noise takes place fast enough for us to do something about it quickly, if we need to. One of the earliest attempts to measure the speed of the nerve impulse involved timing how long it took for someone to react to a touch on the toe, compared with reaction time to a touch on the thigh. The second reaction was quicker than the first, because the nerve signal had less distance to travel to the brain and back.

The difference between the two times showed that, compared with electricity, the nerve message dawdles along its fibre, barely reaching the speed of sound. How fast the message actually travels depends on two things : the thickness of the nerve fibre, and whether or not it is covered with a special coating that speeds up the impulse. The difference these factors can make is considerable: a fast nerve fibre can transmit a message up to 100 metres in one second while a slow one will cover only one metre in the same time. (There are, of course, many variations in the speeds of conduction in the different nerve fibres of our bodies.)

These different speeds of nervous activity can actually be felt when, for example, hot liquid falls on your foot, perhaps from a bath tap. First, the spinal reflex starts to operate before pain is consciously felt. Then you feel two sensations of pain, one rapidly and another, duller sensation that follows up to half a second later. These two signals travel between the same two parts of the body – the foot and the brain – but along two different types of nerves, which are responsible for their different speeds of travel.

Both types of nerve are like a long, hollow tube of jelly but one, the faster, has an extra coating called a myelin sheath. In the slower nerve, the message travels as a steady wave; in the faster nerve, the coating helps the nerve message to travel at up to 100 times the speed of messages in the slow nerve. The myelin sheath consists of a series of cells wrapped round the nerve fibre, along its length, like fried eggs around a rope. There are gaps between each myelin cell and the nerve impulse travels along the fibre by leaping quickly from one exposed section of the fibre to the next.

Most of the body's nerve messages travel along paths made up of several different nerve cells. Each time a message reaches the end of its fibre, it makes contact with the next nerve cell and can trigger it to fire. With the knee-jerk reflex, as we saw, the message travels along set pathways which provide no opportunity for variety – there is only one possible response to a tap on the knee or a stretched thigh muscle. But there are other reflex activities where, although the overall *purpose* is always the same – to escape from pain, for instance – it could be achieved in several different ways, making use of a number of different routes.

The various actions the body may have to perform to escape the painful stimulus are not at all predictable, since they usually depend on the situation. For example, when one leg has to be pulled quickly away from something causing it pain, adjustments must be made to prevent the person overbalancing, muscles in the

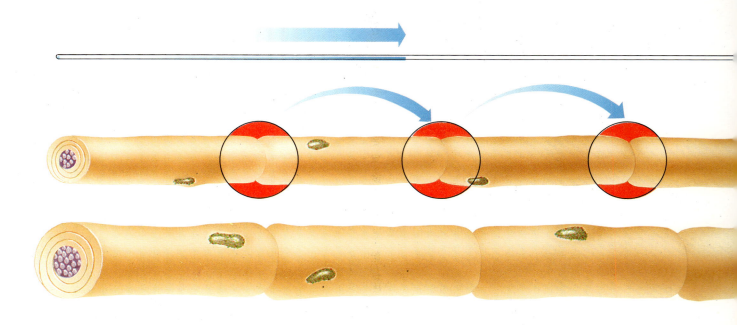

other leg must adjust to take the weight, arms might have to stretch out suddenly for support, vocal muscles may contract to say 'ouch', and the hairs on the legs may be raised with the pain.

All of these actions happen in a fraction of a second and all of them are triggered by one incident – a pain in the foot. Once the pain stimulus arrives at the spinal cord it sets off a cascade of nervous activity along a whole network of nerve cells. Each of the new nerve signals will end up at some muscle that plays a part in the overall pattern.

When danger threatens, the body is programmed to do one thing – remove the threat as quickly as possible. Any hesitation could lead to injury. This is one of the reasons why it is so useful to have decisions made lower down the body than the brain. If the foot is in trouble – if it is trodden on, for example – the brain is nearly 2 metres (6ft) from the action. At a speed of about 20 metres (65ft) per second – far less than the speed of sound – the message could waste valuable time if it had to go to the brain and back.

Throughout the length of the body there are many nerve circuits programmed to respond to some initial event without our conscious control. They take care of the basic activities which all humans need to live a normal life. And each of them, however complex, is built up of the simplest of circuits – a series of electrical impulses and two or three connecting nerve cells.

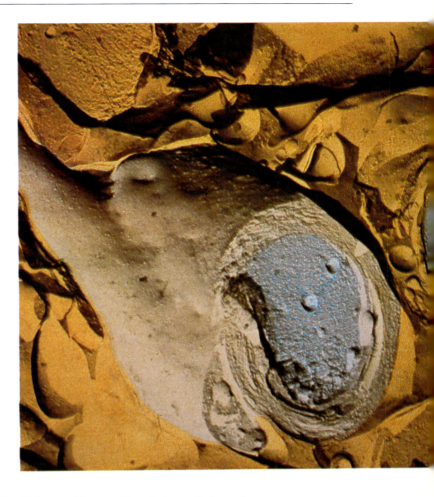

The speed at which a nerve impulse travels depends on two things: on the thickness of the myelin sheathing and the diameter of the nerve or nerve fibre. The thicker they both are, the faster the speed of transmission. Not all nerves or nerve fibres need to be very fast transmitters, of course. In some, impulses travel at a speed of 0.5 metres (1½ft) a second, in others at 120 metres (390ft) a second.

The diagram below shows how, instead of travelling steadily along a fibre, nerve impulses leap from one myelin junction to the next. Note the nuclei of the myelin-producing cells. Transmission would be faster in the bottom nerve – it is stouter and has a thicker coating of myelin.

The photograph above shows a nerve fibre (axon) in cross section. The fibre itself is shaded blue. Wrapped around it, Swiss-roll fashion, is a coat of myelin, a fatty substance which insulates many nerves and nerve fibres in the human body. Myelin is produced by specialized cells which grow around nerve fibres.

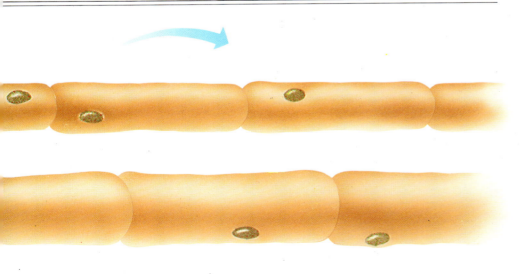

9 · Maps on the Brain

Our lives are full of decisions. Some – a few – make the difference between life and death, but every decision has consequences of some sort, good or bad. Our decisions are reached by our brains on the basis of what we see or hear, or sense in some other way, and they are carried out with our muscles.

The business of living *demands* decisions. In any ordinary day, we make thousands of choices: every changing situation presents us with the choice of what to do next. If we analyse one very simple decision, we can see how many different elements in the body are involved in making sure that the correct choice is made. And 'correct' usually means the one that has the most beneficial outcome.

Imagine a pilot, preparing to take off in his light aircraft. At a crucial point in his take-off, he sees a car crossing the runway in his path. He could do one of two things: stop the plane, and hope to come to a halt before hitting the car, or take off immediately, and hope to avoid hitting the roof of the car with his undercarriage.

As we analyse the events of this split-second decision, we may be able to glimpse how the brain organizes its information flow to and from the world outside. The pilot's *senses* need to take in the surrounding information. His *muscles* put his decisions into effect. Between them, the brain has a very complex task.

At any one time, the brain is a junction box alive with incoming and outgoing messages. These are carried by thousands of millions of cells in the nervous system. From birth and before, these cells grow and develop more and more connections with each other. They form such a complex arrangement in the brain that, at present, we have little chance of understanding what individual cells or clusters of cells do. We cannot tell what specific thoughts or memories they hold but we can discover quite a lot about the way in which they are organized.

We can find very few clues about what really goes on beneath the surface by looking at the shape and surface structure of the brain. It is just a large blob of jelly-like material, with a shiny grey crumpled surface. The individual nerve cells cannot even be seen without a microscope, and the only visible difference between various areas is that some are grey and others are white. The outer layer of the brain is called the cortex, and plays a very large part in our human behaviour.

The bumps and grooves on the surface mean that a much larger area of brain cells can be crammed into a small space. As the mammalian brain evolved its surface became more and more convoluted, leading, in man, to a much greater surface area than that possessed by any other creature of comparable size. If all the crinkles were smoothed out, the cortex would cover the area of a pillowcase, and we would need an impossibly large head to contain it. So evolution has found a way for us to be more intelligent without a larger head.

One of the most obvious things about the shape of the brain is that it is divided into two halves. Like the surface of the body, it is broadly symmetrical. Each half of the brain controls half of the body but, interestingly, the *left* half of the brain controls the *right* half of the body, and vice versa.

The sensory cortex

If we investigate how this control is carried out, we find that on each side of the brain one particular area, the sensory cortex, deals with the sensations experienced in the opposite half of the body. Whenever we feel a sensation – touch, heat or pain – in one part of the body, say the foot, a message travels along a nerve pathway from the foot to one tiny area of the cortex. In fact we only know that the foot is feeling anything when a message arrives at this specific part of the brain.

There is nothing in the nerve message itself that shows where it comes from; all nerve fibres use the same code and say very similar things. The message only makes sense when it arrives at its destination. This is rather like the use of electricity in a house: it is all the same until it arrives at the cooker, the television or the lamp. Then its effects are very different.

When nerve messages arrive at different parts of the cortex, the connections between the brain and the rest of the body form an interesting pattern on the surface of the brain. In fact, the messages from the leg arrive near the messages from the foot, and those from the arm arrive near the messages from the hand. If we plot them all out, we see that this strip of brain contains a map of the human body.

The proportions allocated to different areas of the body reflect the importance the brain gives to different sense messages rather than the spatial relationships of the parts of our bodies. Because our mouths and lips are

Without a cortex the pilot on the right would be able to move in a coordinated manner, tell up from down, eat and sleep regularly, and perhaps grunt if unpleasantly disturbed, but he would be unable to see, smell, or speak, and he would display no interest in or memory of his surroundings. In other words, he would lack most of the abilities we define as intelligence.

The picture of the brain below right shows the mantle of nerve cells that make up the cerebral cortex – cortex means 'rind'. The staining technique has also picked out the many folds of the cerebellum **A**. The cerebral cortex varies in thickness between 1.5 and 4.5mm ($\frac{1}{16}$ and $\frac{1}{5}$in).

The grey appearance of the unstained cortex – and hence our equation of intelligence with 'grey matter' – is due to the density of nerve cell bodies it contains. As the electron micrograph below shows, each nerve cell has connections with many others.

By directly stimulating certain areas of the cerebral cortex with electrodes it has been possible to find out which areas are concerned with moving various parts of the body. By placing electrodes on the scalp it has also been possible to find out which areas of the cortex are most active when we are talking, reading, listening to music, and so on. But there is considerable overlap between our motor areas and our sensory areas. Attributes such as memory, imagination and personality are associated with the frontal lobe of both hemispheres – the cortex that lies directly behind our forehead – but not in any localized fashion.

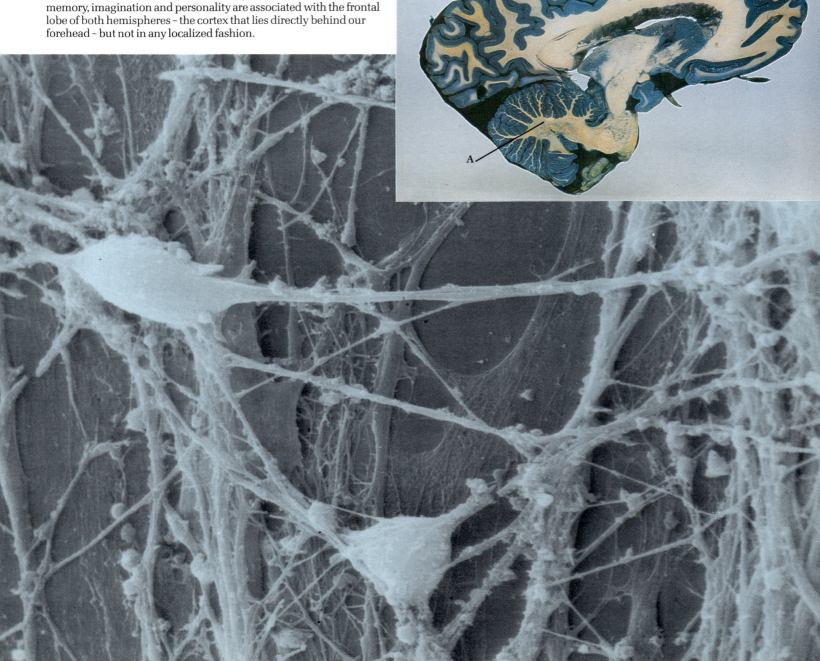

A

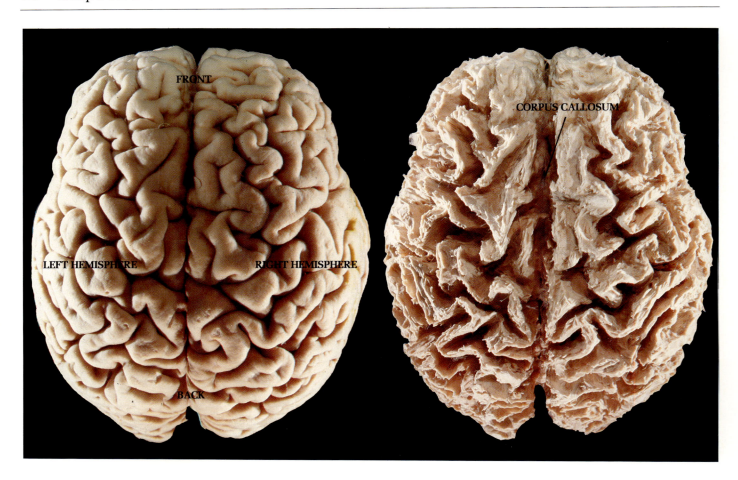

very sensitive, there is a larger area of brain to deal with their sensations. The hands, with their accurate sense of touch, also need a lot of brain cells. But the back and the neck and the top of the head relate to relatively small areas because they seldom send important sensory messages to the brain and so they do not need as many brain cells to deal with them.

The visual cortex

There are some important types of sensory message which are not represented on this particular part of the brain map – the senses of vision, hearing, taste and smell. Vision, in particular, is so important that there is an entirely separate part of the brain devoted to it.

If we return to our imaginary pilot, the vital decision he is about to make depends on the messages his eyes send to his brain. Certainly, his sensory cortex will play *some* part – there will be messages arriving there that carry information about the pressures on his feet and hands, and this will tell him whether the controls will do what he wants. But far more important are the danger signs contained in the visual messages he is receiving. The presence of a car on the runway, its speed, the aircraft's speed judged by the changes in images of the

runway and horizon – his assessment of these will depend on how the brain treats the millions of flickering nerve messages from the two eyes.

On a patch of brain tissue low down at the back of the brain, the cells respond in a very pictorial way to what the pilot sees at any moment. But, as with the other sensory messages, there are certain important distortions. Just as the sensory strip of the brain gave a lot of significance to the hands, the visual cortex devotes a lot of brain cells to images on the tiny area at the centre of the retina. This is where our vision is most accurate and most colourful. So when the scene is laid out on the brain its centre is greatly magnified.

The eyes will automatically move towards the most important part of the scene so that its image falls on the centre of the retina and the brain can deal with it effectively. So long as the pilot's eyes follow the car and keep its image in the centre of the retina, he has many more brain cells to deal with the image of the car on the runway, and of its driver, than with anything on the edge of the airfield.

The cells of the visual cortex are organized in hierarchies. Some of them respond just to light at a certain position in the visual field. But through

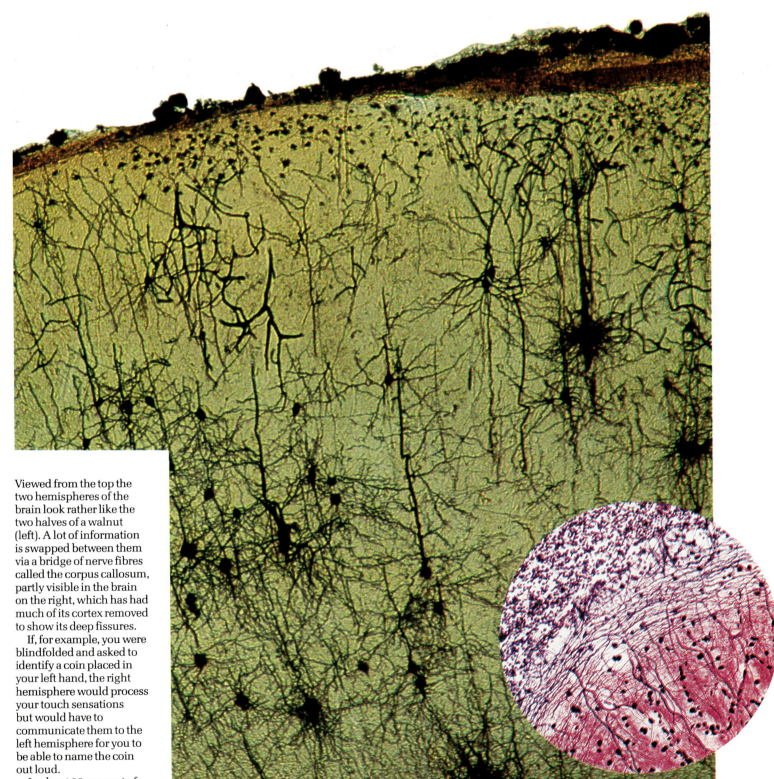

Viewed from the top the two hemispheres of the brain look rather like the two halves of a walnut (left). A lot of information is swapped between them via a bridge of nerve fibres called the corpus callosum, partly visible in the brain on the right, which has had much of its cortex removed to show its deep fissures.

If, for example, you were blindfolded and asked to identify a coin placed in your left hand, the right hemisphere would process your touch sensations but would have to communicate them to the left hemisphere for you to be able to name the coin out loud.

In about 90 per cent of people the left hemisphere is associated with logical, analytical and verbal activity, while the right is concerned with processing spatial and emotional information, and with visual memory.

Nerve cells in the cerebral cortex are classified according to the shape of their cell bodies; the nerve cells most in evidence in the photograph above are pyramidal neurons; these are some of the longest and provide communication between superficial and deeper layers. The dots near the surface are cross-sections of neurons most of whose connections are horizontal. Since only a proportion of nerve cells take up the stain, in reality many more are packed into the tiny area of tissue shown above. The reduced magnification and the staining in the photograph inset show the layered appearance of the cortex more clearly.

elaborate interconnections, other cells respond to specific features of the image, wherever they are in the visual field. So, for example, if several brightness detectors in a straight line all fire at the same time, their links with the next level in the hierarchy will cause one particular cell to fire, thus acting like a straight line detector. With further levels of connection, two straight line detectors firing at the same time could cause an 'angle detector' to fire. In this way, what started as a simple pictorial representation at one level in the brain is broken down and analysed into meaningful elements. There are even brain cells that are fired only by quite complex patterns, if they are important to us.

One very meaningful visual pattern for man is the human face. There is some evidence that the brain contains 'face detector' cells, which fire only when we see a human face. These could work through connections with other cells that respond separately to eyes, nose and mouth. Even newborn babies apparently recognize a human face, so it seems likely that we are born already equipped with this particular piece of nerve circuitry.

Making a decision

By organizing the patterns of light, shade and colour into familiar objects which mean something to us, the visual cortex is sharing some of the mental work involved in making a decision. For example, the pilot sees a succession of red patches as a car heading for the runway – he does not need to work out what it is. Once that idea reaches his brain, it presents him with a split-second choice – to continue the take-off or to abort it. He

The visual cortex, which is situated low down at the back of the cortex, in the occipital lobe, processes information coming from both retinas. Below, in a section through the visual cortex of a monkey, the alternating light and dark bands show the adjacent cell groups that handle messages arriving from the left and right eye.

Some cells in the visual cortex are most strongly stimulated when the messages coming from each eye are similar but not identical. These are probably the cells responsible for our perception of depth and our judgement of distance.

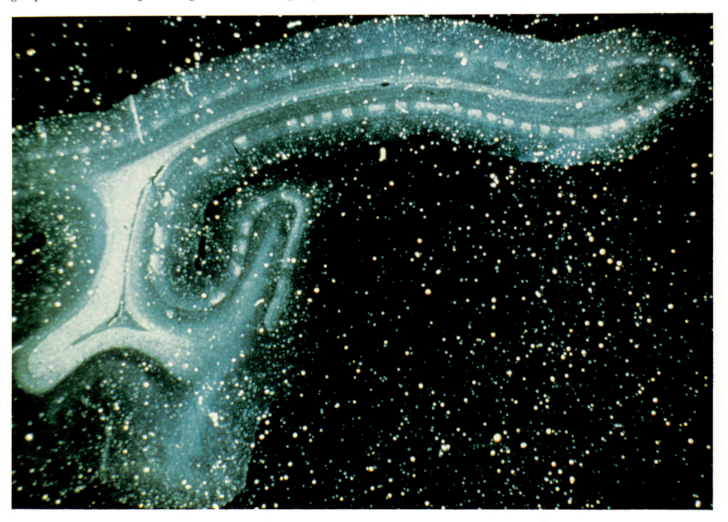

has to choose one of those two actions and then move his hands or his legs so that his decision is carried out.

So far the brain has received all the information it needs to make a decision. It has a picture of the pilot's body represented on the sensory cortex – with information about the position of his hands and feet, the pressures caused by acceleration of the aircraft, and so on. It also has a picture of the outside world on the visual cortex – the car on the runway, the speed of the aircraft, the readings on the dials in the cockpit.

But there is one other factor that could influence the pilot's decision – his memory. Most of our decisions use the stored memories of similar experiences in the past. Have we been in this sort of situation before? What did we do then and what was the result ? Emotion is also important. Do we play safe or do we take risks ? Are we easily panicked or can we keep a cool head ? Memories and emotions lie deeper in the brain, but they have connections with many different parts of the cortex. Images and sensations are stored away as we experience them, together with their emotional overtones. If we see or feel the same way again, the pathways between the cortex and stored experiences will enable us to recognize the familiar sensations. We can also actively seek the consequences of previous actions from where they are stored in our memories.

At the moment of choice, the massages carrying the various pieces of information in the brain converge. The picture of external events and the state of the aircraft meets the pilot's memories and experiences, and he has to weigh up which action to take. At the level of nerve cells, decision-making is really like voting. An

The diagram below attempts to suggest the piecemeal analysis to which arrays of cells in the visual cortex subject a complex shape – such as a car – before we 'recognize' it. Recognition involves the integration of lots of bits of information from different arrays of cells. Some cells in the visual cortex respond to point stimuli, others to horizontal lines, others to vertical, others to lines at certain angles, others to combinations of lines. Recognizing a shape like a circle probably involves the cooperation of a lot of cells, each dealing with lines at a slightly different angle and all firing simultaneously.

individual nerve cell, such as the one leading eventually to the arm muscles, will only fire if it receives enough incoming votes. It is in contact with messages from thousands of other cells, each of which can transmit a small current, or none at all. Some of these cells excite the nerve, others dampen it down. Only when there is a large enough balance of messages that *excite* will the nerve fire, setting in motion the particular muscular movement. So in this case, the nerves that lead to the plane taking off will only fire if there are considerably more 'yesses' than 'nos'.

Although the process is very complex, we can be sure of one thing – if a choice has been made in the past that had a happy result, we are more likely to make a similar choice in the future. In the same way, if a nerve pathway has carried a successful message once, it may actually change physically so that it is a slightly easier pathway for the next nerve message to travel along. And if a message along a certain pathway was followed by an unpleasant result it may be more difficult in future for messages to travel along it.

Acting on a decision: the motor cortex

A decision is no use if it stays in the brain. When we decide to move an arm or a leg, the arm or the leg must be told of this decision. It must also be told how far to move and with what strength.

Of course, with the decision facing the pilot, there is far more than *one* nerve cell involved. It is likely that whole chains of nerves fire, inhibiting or exciting each other at many points of contact within a fraction of a second. But, with a simple choice, there has to be just one result, a nerve message to a muscle. The area of the brain that makes the decision and the area that carries it out are very near to each other. When the pilot decides

whether to take off or brake, the final decision results in a message going out to the muscles from an area of the cortex that initiates all voluntary muscular actions.

Like the sensory cortex, this strip of brain can also be looked at as a map of the body. Here the amount of brain tissue dealing with a particular organ or limb depends on how much we need to *move* it and how accurate the movements have to be. The sensory cortex dealt with incoming messages; the motor cortex – as this area is called – sends messages out. Although its map of the body is similar to the sensory map, the importance placed on the various limbs and organs is different. The hands, for instance, are coordinated by an even bigger area, as the muscle of the hands and fingers can move very accurately and develop a wide range of forces.

If the pilot chooses to brake, he will have to press hard with his feet, and so a message will go out from a point on the motor cortex in the area that deals with the leg muscles. If he chooses to pull back the joystick to make the plane take off, then the message will go to the arm muscles from the brain cells that deal with them, which are located higher up the motor cortex. If his judgement is correct – if his nerve cells have summed up all the incoming messages correctly – the aircraft will avoid the car and the pilot survive to fly another day.

All this is just one decision among thousands the pilot will make in a day. His brain reached the decision by analysing millions of bits of incoming information and consulting a bank containing millions of memories. It then sent out several instructions to various parts of the body, and he avoided danger.

In order to be able to do all this, as we have seen, the brain actually contains maps of the important aspects of our personal world, and makes its decisions by connecting up one brain map with another.

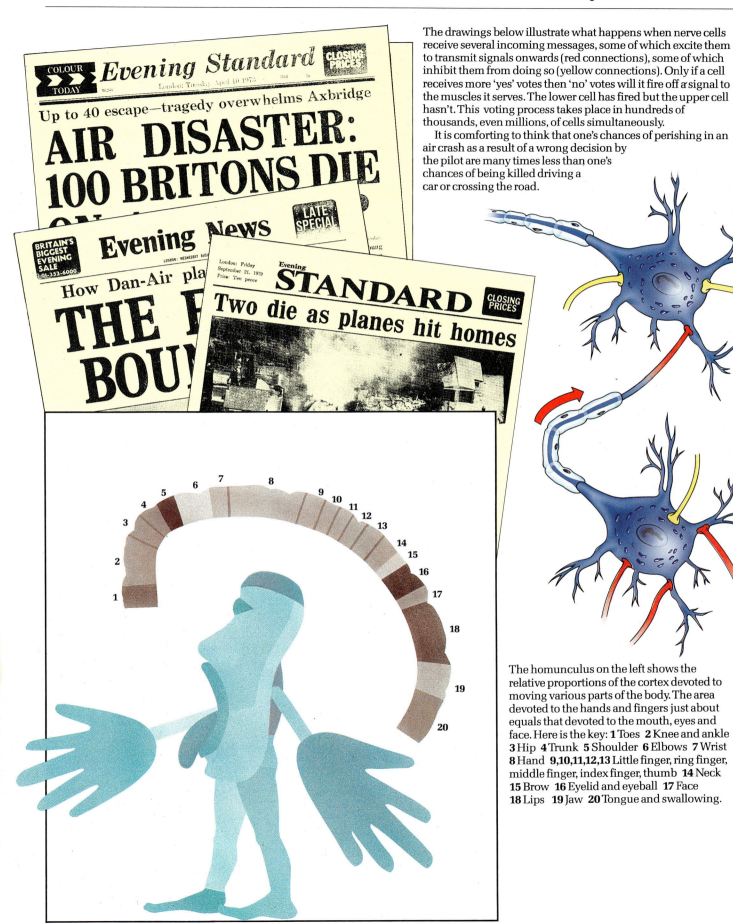

The drawings below illustrate what happens when nerve cells receive several incoming messages, some of which excite them to transmit signals onwards (red connections), some of which inhibit them from doing so (yellow connections). Only if a cell receives more 'yes' votes then 'no' votes will it fire off a signal to the muscles it serves. The lower cell has fired but the upper cell hasn't. This voting process takes place in hundreds of thousands, even millions, of cells simultaneously.

It is comforting to think that one's chances of perishing in an air crash as a result of a wrong decision by the pilot are many times less than one's chances of being killed driving a car or crossing the road.

The homunculus on the left shows the relative proportions of the cortex devoted to moving various parts of the body. The area devoted to the hands and fingers just about equals that devoted to the mouth, eyes and face. Here is the key: **1** Toes **2** Knee and ankle **3** Hip **4** Trunk **5** Shoulder **6** Elbows **7** Wrist **8** Hand **9,10,11,12,13** Little finger, ring finger, middle finger, index finger, thumb **14** Neck **15** Brow **16** Eyelid and eyeball **17** Face **18** Lips **19** Jaw **20** Tongue and swallowing.

10 · Consciously Human

Most of the activities of the human body have their counterparts in other animals. But there is one area in which humans are completely different, and that is thinking. Human life is richer and more varied than the behaviour of animals in our shared world, and much of this difference springs from the human ability to memorize, use language and teach each other.

There is no visible reason why these differences should be so dramatic. Individual nerve cells in the brain work in the same way whether in a human or a dog and in many other ways our nervous systems are similar to those of other animals. We all have similar reflexes, for example, which work through the spinal cord. And we all have areas of the brain that receive our sensations and give orders for our movements. But we differ from other animals because of the way our brain cells *combine* their activities to achieve astonishing feats of complexity that are beyond any other animal's brain.

Both humans and animals need to find food from time to time; they need to be able to recognize danger so as to avoid it; and, for the sake of the species, they need to be able to recognize potential partners for reproducing. To do all these things, we need a brain and a nervous system to take in information and to decide whether to act on it or not. We do not eat whenever we see or smell food, only when we require it. As we walk through the rich world of sense messages, our brains give those messages meaning. But when you take your dog for a walk you will see significance in sights and sounds that have no meaning for your dog, while he may become very excited about smells which you hardly ever notice.

These differences are most obvious when it comes to language. We will see how this uniquely human activity is embedded in the structure of the human brain, in a way that has made all the intellectual achievements of our species possible

On the top surface of the cortex are three of the most important sensory areas. As described in the last chapter, these are the brain cells that receive and deal with messages from the organs of vision, hearing and touch. In another part of the cortex, on the underside of the brain, are areas that deal with more primitive sensations, such as smell, hunger, thirst, fear, sexual feelings and aggression. These areas are sometimes called the limbic system, and one of the most important differences between our brains and those of most

animals lies in the way these various areas connect up with each other. To understand how our brains work differently we have to look at the way we make connections in our minds, in other words how we learn.

With some tasks, we probably learn in a similar way to animals – learning where to find food for example. When we see food, and are hungry, we are likely to eat it. There is clearly a link in the brain between the 'limbic' sensation of taste or satiety and the part of the brain that deals with vision. Both humans and animals are likely to remember this experience and seek the same situation next time they are hungry.

It is even possible for animals to learn *indirect* associations that lead to a rewarding situation, as well as the direct sight or smell of feed. As Pavlov discovered, animals can associate the idea of food with irrelevant sights and sounds. If a particular symbol is followed often enough by food and the 'reward' of satisfying hunger, it will *mean* food, and can even produce physical responses such as salivation in anticipation of eating. When this happens, the brain forms connections between non-limbic sensations, such as sights and sounds, and the more primitive sensations of hunger or thirst or fear. But there is another type of learning that animals find much more difficult; this consists of making connections between pairs of sensations such as sight-sound, sound-touch, touch-sight and so on. This is probably because they have few pathways in their brains between the sensory areas that deal with these sensations.

Associations in the brain

In many animals the sensory areas dealing with sight, hearing and touch take up most of the cortex, but as we look at more and more advanced animals, going, for instance, from dog to primate, we find that these zones become increasingly separated by other areas of brain tissue. By the time we get to ourselves, the brain has small strips of tissue which we know receive sensations or give orders, surrounded by larger areas which are connected to them. These areas are called the association areas of the cortex.

Without large association areas on the cortex animal thought seems to use a comparatively simple set of pathways in the brain, connecting the sensory part of the brain with the basic animal drives. This simplicity is

The bodies of humans and animals have far more similarities than differences, particularly when it comes to the structural and physiological details of individual organs and systems. In fact most mammalian brain cells – those of a dog or a cat, say – are indistinguishable from those in a human brain. Yet we use our brains to do things that animals cannot do. We play musical instruments, we draw portraits of each other, we appreciate abstract shapes and arrangements of colour. So wherein lies the difference? Probably in the way our brain cells connect with each other.

The connections between groups of cells in the human brain may actually be programmed in our genes. In this way human infants are predisposed to learn the sounds of human speech, predisposed to paint figures without bodies before they paint figures with legs and arms *and* bodies, predisposed to understand that objects continue to exist even if they cannot be seen, and so on. In other words, our brain circuitry equips us to learn and use symbols.

Inside the human brain are a series of paired interconnecting chambers or ventricles filled with cerebrospinal fluid, the same fluid that cushions the spinal cord in the vertebral canal. They are seen here from the side and from the top. The cerebrospinal fluid is secreted by specially adapted blood vessels in the roof of the ventricles.

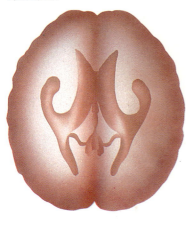

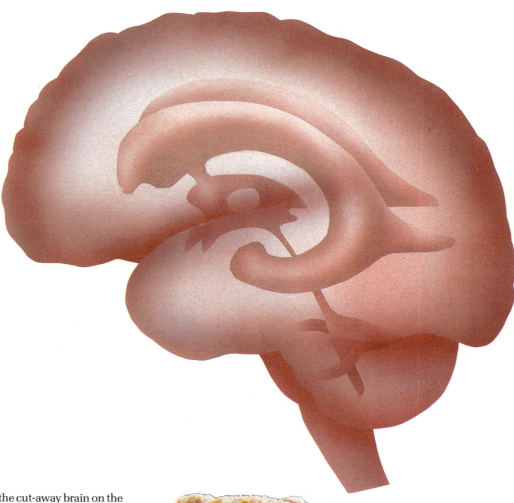

The bundles of nerve fibres shown in the cut-away brain on the right link areas of the cortex. These links allow us to 'see' someone when we hear their voice or gain an idea of an object through touch alone. Sensory information in the cortex does not have to be referred elsewhere for these associations to take place.

Shown separately below is the club-shaped midbrain and brain stem. The club part is the thalamus, a complex relay station. The stalk part contains, among other things, the medulla oblongata, which controls breathing and heart beat, and the reticular activating system, which keeps us awake and sends us to sleep.

reflected in the bundles of nerve fibres in the animal brain which, on the whole, travel mainly between these three areas – the sensory cortex, the limbic system and the motor cortex. But in humans a lot of learning makes use of the association areas without ever involving the more primitive sensations. Each specialized zone – the visual cortex, for example – only connects with the neighbouring area of association cortex. *Those* areas can then send and receive messages along many nerve fibres which connect them to many different parts of the brain, including the limbic system.

Because of this, we can make quite simple associations of ideas that are entirely beyond the abilities of animals. If we see a shape, and then try to detect a similar shape by touch alone, we do not find it difficult. We perceive that the sight and the feel are part of the same object. A coin that we see in our hand is worth the same as a disc that we feel in our pocket. Our senses of sight and touch have met up in one of the association areas and agreed with each other.

In fact, we can link up any two senses quite easily, and this is why we can also connect the sight of a shape with the sound of the word describing it and even the written word, and see them as all meaning the same thing. But when an animal sees a square, the sight of it can only link up with the deeper parts of his brain – he has no connecting nerve fibres that would enable him to link the *sight* of a square directly with the *sound* of the word. If he does learn to make the connection it is probably via one of the deeper centres – if food is given as a reward, for example.

The Russian physiologist Ivan Pavlov (below right) trained dogs to salivate when a bell sounded – after learning the bell-food association, they salivated in response to the bell alone.

Of course, when we first started seeing and feeling shapes, we did not make the connection, but because of the existing pathways in our brain we were able to learn very easily what seeing and feeling had in common.

Memory

Developing these many links between different parts of the cortex puts a great strain on the capacity of the brain. To make links between, say, sights and sounds and sights and touch, in addition to sight and hunger, or thirst, means that there have to be three or four times as many brain cells to cope with the increased flow of messages between different parts of the brain.

This increase in connections has been accompanied by a need for more storage space. For the most significant thing about this newly evolved associative ability is that it allows us to use symbols – sights and sounds can stand for other, more primitive sensations. Symbols can mean food or danger or sex, and this multiplies the number of different ways of storing the same idea in the brain. We could only really exploit our

A dog can make the connection between food and a special 'food' tone of voice on the part of its owner because this involves the swapping of information not in the cortex but in the limbic system. The limbic system, which is buried in the core of the brain, integrates sensations like hunger with long- and short-term learning.

ability to make symbolic associations by developing a very good information store – our memory.

We use memory in many different ways. In some situations, it is useful enough for us to remember over quite a short period. Many sights and sounds stay in the brain for half an hour or more without us trying to remember them.

It is not known *how* this short-term memory is stored, but we have some clues as to *where*. It seems that experiences that are sensed through the eyes are stored near the visual cortex, things we hear are stored near the part of the brain that deals with hearing, and so on. But many experiences come in as mixtures of sensations. When we observe a complex part of our human world – a motor car, for instance – we remember it as one unified object rather than as several different sensations of sight, sound, smell and so on. Somewhere in the brain is a trace of that car. And it is a trace that can be recalled in greater or lesser detail for some time afterwards. So how is it stored?

One possibility is that some memories are stored as currents which flow round and round a small circuit of nerve cells. The current could be quite strong at first, but would then become weaker and weaker until we could no longer detect it. This theory is supported by experiments which have found that if a small patch of brain cells is stimulated by one burst of electrical activity, the nerve cells in it can continue firing for up to 30 minutes.

Many of our memories seem to disappear after a few minutes, but some of them remain and find a more permanent place in the brain. It is thought that long-term memory may involve structural or molecular changes in the brain cells, although we are still a long way from discovering the true picture. Although remembering is not a uniquely human ability, we probably hold more information in our memories and can retrieve more than any other animal. And we use it for a wider range of activities.

It sometimes seems as if there is no limit to the

In addition to long-term memory, we have what scientists call 'working memory'. This is the brief-trace memory that enables us to do arithmetic and to listen to the end of a sentence and still remember the beginning. Short-term memory relies a lot on 'echoic' memory – that is why we mutter telephone numbers to ourselves while we dial them – and on what a leading memory scientist calls a 'visuospatial scratch pad', the ability to project scenes, objects and numbers onto an imaginary screen in front of us.

contents of an average human memory. Yet how can slightly more than a kilogram of tissue hold the experiences of a lifetime? The human brain stores as much information as a well stocked library, and we do not just store our memories at random – we can actually find them again when we need to. If we want a book from the library, it is not enough to know that the book is there somewhere; a store of printed memories is no use unless there is a system to locate the information we want at the time we want it. If library books were stuffed into the shelves at random, the library might be an interesting place to browse in, but would be useless as a tool for living.

The clue to the storage and retrieval of memories lies in the plentiful supply of brain cells. If you count all parts of the brain above the brainstem, there are probably 100 000 million brain cells, and even in the cortex, the thinking part, alone there are probably about 10 000 million.

Any one brain cell is capable of quite complex activity on its own. A single brain cell has many projections from its cell body: there are shorter projections called dendrites which receive messages from other brain cells, and there is usually a longer projection called an axon which passes on signals to the dendrites or cell bodies of other brain cells. It is really just a specialized link in a chain that can receive messages and sometimes pass them on. Through its dendrites and its axon, the average brain cell links up with about 60 000 others; indeed some cells have links with up to a *quarter of a million* others.

One way our memories could be stored is as individual pathways among some of these cells. A unique memory could then be stored in the form of a chain of nerve cells, with the links being a few of the thousands of points of contact between each cell in the chain. Even a few cells connected with each other allow a lot of different pathways. With 10 000 million cells to play with, and each of them connected with many thousands of others, it would be possible to remember a

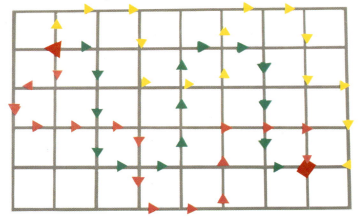

If memories are stored as pathways between cells, there is no shortage of pathways between the thousands of millions of cells in the human brain. Even a simplified two-dimensional network like the one above, with only a few locations with only four connections each, provides many different pathways (yellow, blue and pink arrows) between two points (the red triangle and square).

Unlike a computer, which will accept and work with all the information stored on the tapes in the picture on the left, the human brain actively selects the information it will work with, put into store and retrieve. On the whole we tend to commit to and retrieve from memory things that are meaningful and needful. Things that do not fit into any of the thousands of overlapping and changing networks of concepts we all have in our heads, we reject. If we are not interested or not paying attention we are unlikely to process information deeply enough to remember it. And some memories are so painful that we put a block on their retrieval – 'repression' psychologists call it.

Unfortunately the proposition that we forget nothing is untestable. But research on the kind of forgetting that occurs in the laboratory and in other controlled situations suggests that it is more a question of recent memory traces interfering with earlier ones rather than the earlier ones fading. Time *per se* does not seem to erase memories.

The yellow patch in the brain scan above shows the area of the left hemisphere most active during speech. In the diagram on the right, which shows the brain from below, with the optic nerves at the top, the relative proportions of the left and right hemispheres devoted to speech are shown in red. More than 10 per cent of left-handed people have speech primarily in their right hemisphere, as do about 2 per cent of right-handers.

Below, for comparison with the human brain, are the brains of a sheep **A**, a cat **B**, a rabbit **C**, a rat **D** and a pigeon **E**. The human cerebellum is completely overhung by the two hemispheres, a consequence of our erect stance. We owe our brainpower not merely to the volume of our two hemispheres but to their many folds and fissures.

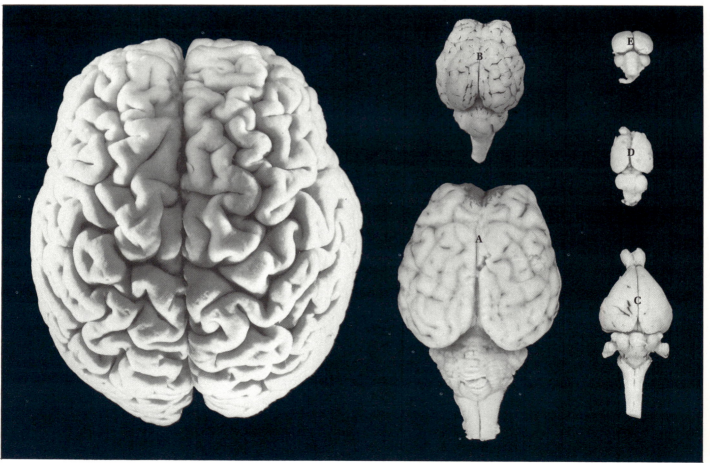

great deal. It has been calculated that the human brain could hold at least *1000 times* as much information in the pathways connecting its nerve cells as is contained in the largest encyclopedia – say 20 or 30 big volumes.

If our permanent memories are stored as connected pathways in the brain, then to remember something we would just have to send a nerve impulse along the same path, 'down memory lane'.

In fact most of our memories are never brought into the conscious mind; we only ever recall a tiny amount of the brain's total store. And yet we can *recognize* images and sounds that we have not experienced for years and years. Faces and songs of childhood; people and places from our teens; books we read, pictures we saw and even smells and tastes from way back in the past – all of these are coded in the human brain, so that we recognize them as 'familiar' if we ever perceive them again.

Many of the things humans choose to remember would probably be meaningless to animals. Of course, animals have long-term and short-term memories as well. But there is something different about *what* we choose to remember – the things we can become interested in – that sets us apart from animals. We saw earlier how human and animal brains might acquire memories that are connected with basic drives, such as the sight and smell of food and water. But much of the richness of human life comes because we do not spend all our time thinking about basic physical needs like eating and drinking and survival. Many of our specifically human activities depend on language and other systems of symbols.

Language

The ability to generalize – to see the symbolic value of images and shapes, for example – reaches its peak of achievement in our use of language.

To an animal, one of the most puzzling human activities must be language. Many of us spend most of our waking hours talking, listening, reading or writing, with no visible reward or apparently useful consequences. And yet language is the crucial difference between humans and animals – it makes possible all our social and cultural achievements. We can store and extract the most complex ideas in the form of the sounds and symbols of human language, in a way that no animal could comprehend. We now know that this is not just a trivial improvement on other animals' abilities to make noises – it is the fundamental property that makes humans human, and it is reflected in major differences in brain structure.

Many of the areas of the human brain that still defy understanding have proven connections with language. A stroke or other injury in one specific area can produce a stream of speech that sounds realistic but has no meaning. The patient has lost the memory for what words go with what ideas. When brain damage affects language it is almost always because the damage has taken place in the left hemisphere of the brain, which controls all speech. The rest of the brain's activities are fairly evenly shared, so that the two halves of the body are controlled by the two halves of the brain. But with speech, it is really all controlled from the left side of the brain. What is more, we can actually see this difference when the brain is dissected: most human brains have an area relating to speech in the left hemisphere much more developed than the corresponding area in the right, whereas animal brains are broadly symmetrical.

It seems that the ability to make all sorts of cross-connections in the brain went hand in hand with the development of language, both spoken and written. With any part of the brain having a connection with any other, we were able to use many of the nerve pathways constructed through experience in the brain to link the sounds of speech with the sight of letters and numbers and give them meanings that were drawn from our memories and feelings. Even hearing a simple question and forming an answer can involve almost every part of the brain. The connections enable us to understand the words as well as just hearing them; to formulate an answer, with the help of our memories; and to send instructions for the right muscular movements to be made by our mouths and vocal cords.

Once we acquired the ability to use symbols, we were able to manipulate words and concepts in our heads, without having ever experienced them in the world.

Now we could consider possible future actions and predict what might happen – a very useful ability in a fast-changing world. Instead of having to learn by experience, we can represent the world to ourselves in our thoughts, think through the consequences of various possible actions, and even benefit from other people's experience by communicating with them.

In this way, many of our uniquely human activities can be seen as a rehearsal for life, and a short-cut to surviving longer. Fiction and drama provide experiences in a far wider range of situations than most of us will ever meet; games imitate mortal combat; it has even been suggested that music trains the mind in anticipating and predicting what will come next, through its inner structures and symmetries; and all of these depend on the ability of the brain to think beyond the sensory messages they convey – to make the right connections.

11 · Muscle Power

However clever we are, we are useless unless we can put our thoughts into action. To live and survive in the world, the living body must be able to act on the world with physical force – and that means with muscles. If animals had not developed muscles, we would still be at the stage of barnacles – sitting and waiting for the world to change around us. With muscular force, we can move ourselves about and find the best environment to live in. We can push hostile objects away and bring desirable things nearer. And muscles that most of us are unaware of control day-to-day events within the body.

'Muscle power' often summons up the image of great force. And indeed, some men with well developed muscles can perform feats of great strength, while others can only look on in admiration. Yet muscles make up 30 per cent of every human body and we *all* use them in ways that demonstrate what a very versatile tissue muscle is.

Any bodily task that requires movement, force or the steady control of a position depends on muscle tissue

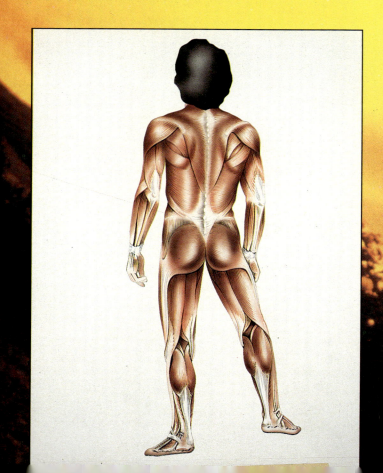

somewhere in the body. A simple action such as taking a drink needs arm muscles to move the drink towards the mouth, throat muscles to swallow, gut muscles to help the drink on its way through the body, and bladder muscles to expel it. Indeed, when we are not moving at all, perhaps slumped in a chair reading a book, muscle power is being generated to hold the body in a relatively alert and upright position.

The muscles that generate enough force to send someone flying and those that perform delicate movements of the fingers operate by the same simple method of contraction, arrived at over many millions of years of evolution.

The muscles in the body can be divided into two groups – voluntary, which we control consciously, and involuntary, which work entirely on their own. In certain circumstances, however, some involuntary muscles may be controlled voluntarily.

Voluntary muscles

There are 600 individual voluntary muscles in the human body and, when we need to, even the least fit of us can generate several kilograms of force across each square centimetre (2-3lb per square inch) of our voluntary muscles. With those at the back of the thigh, we can generate a tension of up to 1200 kilograms (2600lb). This means that if all the muscles in our body could all pull together at the same time, they could lift 25 tonnes (tons). Regular use of muscles can greatly increase their power, and the voluntary muscles of a Kung Fu expert, for instance, may be able to deliver three times the force that most of us can manage.

Although we tend to think of muscles in terms of
strength, many of the muscles in our body exert
very little force – the tiny muscles in our eyes and
ears, for example, make extremely delicate
adjustments.

Well over half the muscles in our body are
inactive at any given moment. That is because
many of them work in pairs or in complementary
groups – one set of muscles turns the head or
flattens the lens of the eye, and another set turns the
head the other way or relaxes and makes the eye
lens fatter.

Most of the voluntary muscles in our body are
attached to two or more bones, against which they
exert their pull. Shown opposite are the superficial
voluntary muscles of the back.

Where does all this force come from? How can a soft, flaccid tissue such as the biceps muscle suddenly generate many kilograms of tension and sustain its force?

Like most voluntary muscles, the biceps is attached at both ends to bones. Its job is to pull two parts of the body nearer to each other by contracting. Tough inelastic tendons connect the muscle to the shoulder blade at one end and to the forearm at the other. Between the tendons is a mass of muscle tissue made up of thousands of parallel fibres. Each fibre is a single elongated cell. Contraction is initiated by a nerve signal from the brain, which instructs a number of the fibres to shorten by a tiny amount; the combined effect is the shortening of the whole muscle.

The instruction travels down the nerve to specialized endings on the muscle itself. These endings appear to be stuck onto the muscle but in fact they are not in contact with it at all: there is a tiny gap between the nerve and the muscle, which the signal has to cross. To produce a contraction, the signal must be ferried across this gap by molecules of a chemical called acetylcholine (ACh). When the ACh arrives at the muscle, it fits like a key into a lock and initiates an electrical impulse which causes the muscle to contract.

Having transmitted one impulse, the ACh which is on the muscle membrane blocks any further impulses. To clear the blockage, another chemical breaks down the ACh so that the muscle is ready to begin the cycle again. Usually, muscle fibres contract many times a second as a succession of chemical messengers pour across the gap and are wiped away.

Just as a muscle shortens by the combined efforts of many individual fibres, so the mechanism by which each fibre shortens can be traced back to microscopic elements working together inside it. Every single voluntary muscle fibre has a pattern of parallel lines at right angles to its length; this appearance is produced by overlapping strands of two long molecules, actin and myosin. For this reason voluntary muscle is also called striated muscle. Muscle contraction really involves interaction between these two molecules; when they telescope together the fibre is fully contracted. If all the fibres shorten at the same time, the combination of these tiny molecular movements can produce an overall contraction of perhaps 50 per cent in length in a muscle like the biceps and a powerful movement of the arm.

So every bodily movement we see is brought about by thousands of muscle fibres and millions of molecules working together, under instructions received down nerve fibres. In muscles that control more delicate movements, such as eye movements, one nerve controls only a few fibres; this arrangement produces

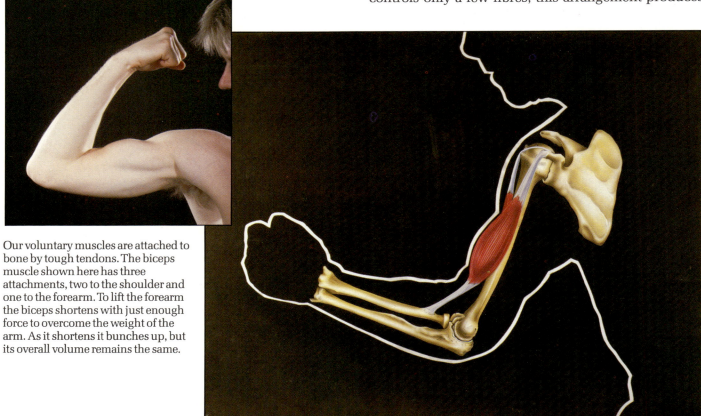

Our voluntary muscles are attached to bone by tough tendons. The biceps muscle shown here has three attachments, two to the shoulder and one to the forearm. To lift the forearm the biceps shortens with just enough force to overcome the weight of the arm. As it shortens it bunches up, but its overall volume remains the same.

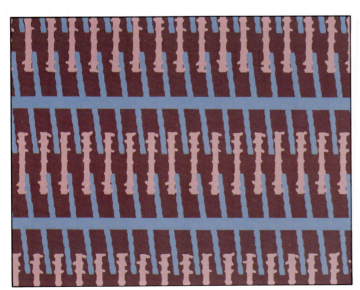

All muscles contract in reponse to nervous signals. Above is the microscope's view of several junctions between nerve endings and muscle fibres; on the right is an artist's view of the same thing. When a nerve message arrives at one of the special junctions it causes an electrical change in the membrane of the adjacent muscle fibre, which then contracts.

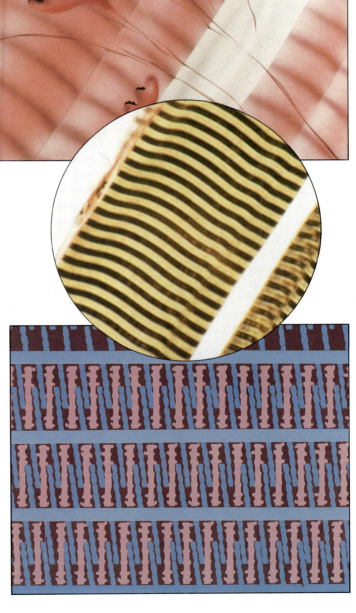

Muscle fibres contract because they contain interlocking strands of actin and myosin which telescope tightly into each other when electrically stimulated.

In the two diagrams below, actin strands are blue and myosin strands pink. On the left they are shown relaxed, on the right contracted. The banded appearance of voluntary muscle fibres – the bands run across the fibres, not along them – is due to the interlocking pattern of these two molecules. The circular photograph on the right shows how these bands look in real life. The bands move closer when the fibre contracts.

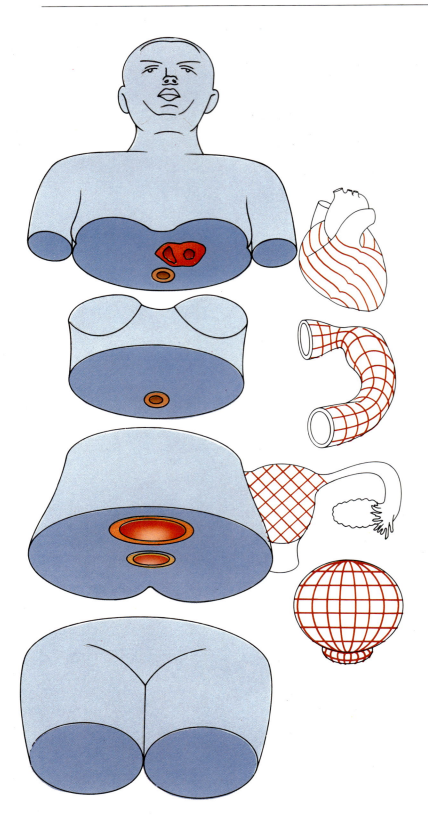

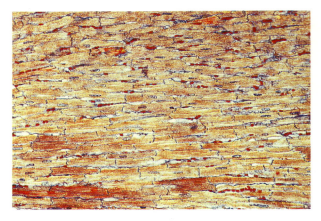

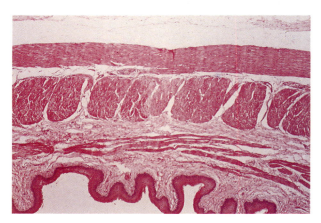

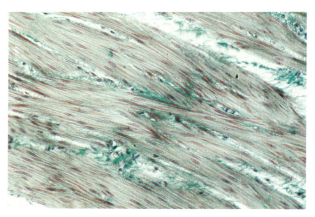

Many of our involuntary muscular activities – the regular pumping of our heart, the peristaltic contractions that squeeze food through the gut, the movements of the uterus during menstruation or childbirth, the expulsion of urine from the bladder – are carried out by muscle fibres that contain only a fraction of the actin and myosin that voluntary muscle fibres contain. That is why they look less stripey – in fact they are often referred to as 'smooth' muscle. The drawings above left show the lie of the main bands of muscle fibres in the heart, intestine, uterus and bladder.

The photographs show, from top to bottom: heart muscle cells; a section through the wall of the intestine, with longitudinal and circular bundles of muscle fibres clearly visible; and muscle cells – in longitudinal and cross-section – in the walls of the bladder.

very fine, graded movements. With bigger muscles designed to make larger movements, one nerve may control a hundred or more fibres.

Muscular force is not always used to produce movements; it may sometimes be used to keep a part of the body in one place. In such cases, muscles stay the same length but develop increasing force in order to prevent movement. These two types of muscle activity are known as isometric ('same length') and isotonic ('same force') contraction. We all use both types of contraction in many everyday activities. When walking, for example, we move our legs isotonically and hold the top part of our body upright isometrically.

Voluntary movements, controlled by our conscious mind, are essential for us to be able to respond to all the unpredictable events of human life. But there are some functions of the body which need movement and force all the time in a very predictable way. Breathing, for instance, requires the diaphragm muscle to move up and down all day and all night at a uniform rate to keep air flowing through the lungs, and this is controlled involuntarily. When we need to override the system, however, which happens when we take a deeper breath, or defecate, or speak, we can move the diaphragm muscle voluntarily.

Involuntary muscles

Many of the body's muscles, in contrast, are totally beyond our voluntary control. The circulation of the blood, for example, depends on involuntary movement in the heart and blood vessels; and the digestive system must always be active on its own. These involuntary systems use a different type of muscle, which can do a great deal on its own.

Muscle cells are at their most independent in the heart. Here it is vital that movement and force never stop for more than a few seconds. Even a single heart muscle cell can beat on its own until it runs out of food or oxygen. When it is in contact with another cell, because of the specialized junctions between them the two will beat time together, sharing a common rhythm; when a whole collection of heart muscle cells are in contact they can start to generate regular pulses of force. To pump blood round the body in a human lifetime, 3000 million contractions of the heart will be needed, and each one starts with the semi-autonomous action of heart cells.

Another type of involuntary muscle lines the digestive tract. This moves restlessly day and night in a never-ending series of waves of contractions that function as a conveyor belt. The smooth muscle that forms the walls of the gut responds automatically to the pressure of the food we have eaten with regular squeezing movements that are a property of the muscle itself. If the wall of the gut is stretched, it automatically squeezes tight, producing travelling waves of force in one direction. Two bands of muscles, one along the gut and one in rings around it, work together to produce this rhythmic rippling, which is called peristalsis.

Smooth muscle also helps us to control when and where we urinate without having to think about it all the time. As we described in Chapter 6, the smooth muscle in the wall of the bladder has an ability to adjust its own tension as the bladder fills, and to keep reminding us to urinate without forcing us to do so.

Energy for muscles

The difference between the weakest and the strongest of us is impressive, but the potential is in everybody to increase their strength because of the way muscle cells work. Each of us is born with a fixed number of voluntary muscle cells which stays the same throughout our lives. But we can generate more muscle power by developing each of the muscle cells we have rather than acquiring new ones. The strength of a muscle cell is proportional to the number of actin and myosin strands it contains. As we grow and become stronger, individual muscle cells increase in size by cramming in more strands. The strongest muscles have very large cells that can generate more force.

If we use one muscle a lot, its cells grow bigger and so it can contract more strongly. This increase in size can be quite localized – one arm that is used more than the other will become much more muscular.

To keep up any sort of muscular activity, muscle cells need fuel and they need oxygen to burn the fuel and supply the energy. Training can also help with this: with practice, we can increase the rate of blood flow to our muscles by up to 30 times. The combination of stronger heartbeat and faster heart rate pushes blood round the body much faster and steps up fuel supplies.

The fuel comes mainly from carbohydrates in the diet. The body stores it as a form of glucose, some in the liver and some in the muscles themselves. We dip into this store continually, taking glucose away in the bloodstream to be burnt with oxygen in the muscles.

This burning of glucose in the muscles makes a major contribution to the body's heat production. Muscle activity produces some heat during contraction, but more as the muscle recovers and prepares to contract again. The burning process also requires more oxygen, and at the peak of activity and for a short recovery period afterwards, heart and lungs have to work harder so that more oxygen can be delivered to the muscles.

In fact, following violent physical activity we go on

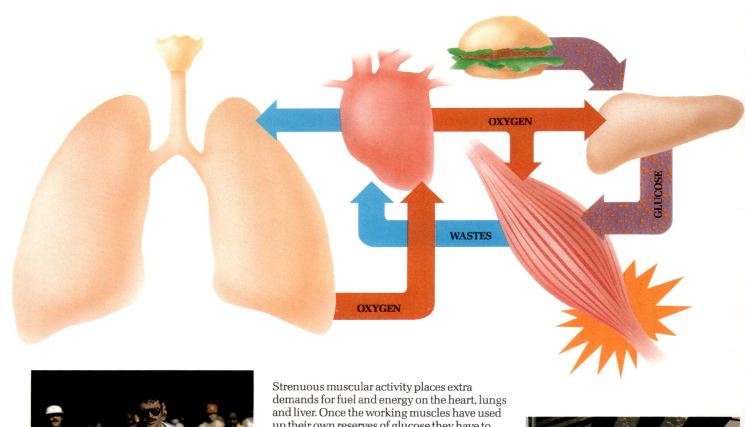

Strenuous muscular activity places extra demands for fuel and energy on the heart, lungs and liver. Once the working muscles have used up their own reserves of glucose they have to depend on glucose released from the liver – the liver stores glucose, derived from our food, in the compact form of glycogen. The extra oxygen the muscles need is supplied by the heart and lungs working harder; the increased blood flow also carries away the extra wastes they create.

If exercise continues, the oxygen delivered by the lungs and heart may be insufficient to 'burn' the glucose the liver has sent out to the muscles, and an 'oxygen debt' builds up. It has to be paid back, by panting, when exercise stops. 'Getting one's breath back' is a sign that oxygen and glucose are once again in balance.

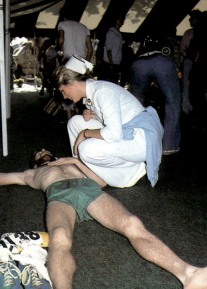

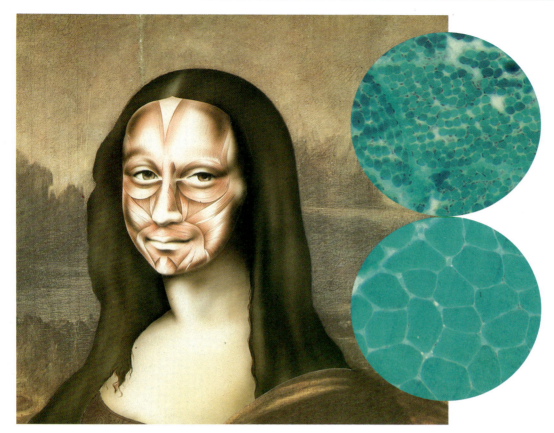

We are all born with a fixed number of muscle cells, and however athletic or sedentary we are that number does not change. But the *size* of muscle cells can change. With a lot of use the number of actin and myosin molecules in them increases. Conversely, muscles that are not used tend to waste away. The quadriceps in the thigh wastes very quickly, for example, if the leg is immobilized in plaster.

On the left is a cross-section through the muscle fibres of a baby – they are thin, unused to exerting force; below are the muscle fibres of an adult.

The face alone has more than 100 different muscles. Though most of them are under voluntary control, many are attached not to bone but to other facial muscles and to the skin. Wrinkles usually occur at right angles to them.

breathing heavily — panting — after the need for energy is over. During exercise we can burn more fuel than the oxygen supply allows, even with maximum heart and lung activity, and can therefore build up what is known as an 'oxygen debt'. When the activity is over, we continue to pant until that debt has been repaid.

Between them, the voluntary and the involuntary muscles have made human survival possible. In fact, nowadays many of these muscles contribute more to our human characteristics than to our survival. In the head alone we find a whole range of muscles, both strong and delicate, which we have come to use in a wide variety of ways.

One of the strongest muscles in the body is the masseter muscle, which raises the jaw and helps it to clamp shut with considerable force to crack nuts or chew meat. Other gentler muscles in the neck have to maintain a steady tension all the time and keep the head upright. The whole rich repertoire of human facial expression is merely an interplay of muscles which evolved for many different purposes. The pupils constrict to improve vision, using a ring of tiny muscles around the pupil. The muscles around the mouth, face

and throat evolved to help us to eat and breathe, but they also allow us to speak and smile. We now use our brow muscles more to express ourselves than to cut down on the light or frighten away predators, the purposes for which they may have originally developed. And the sensitive muscles of the lips can be used for such unimportant tasks as whistling and kissing.

All human skills depend on the remarkable properties of muscles. Any human designer might find it a daunting task to create a device that would be capable of all the activities of muscle. Some muscles can lift hundreds of kilograms or pounds; others only squeeze tiny amounts of blood through the capillaries. Tiny muscles in the eye react with lightning speed, while others in the intestine are sluggish and slow to respond. Some are linked with the nervous system and can only respond to specific instructions from the brain, while others act independently for a whole human lifetime.

When the whole body has to move, many millions of muscle cells work together, in an act of collaboration that shares the effort among muscles widely dispersed through the body. We will see in the next chapter how this coordinated control of movement comes about.

12 · Moving Parts

Muscles provide the power for movement, but that power has to be harnessed and controlled to produce the wide variety of our physical activities. And that variety also depends on the arrangement of the body's movable parts – principally the limbs and trunk, but also smaller elements such as fingers and toes. The ways in which all these elements are linked with each other predetermine the actual movements we find ourselves able to make.

The starting point for any movement is a nerve signal. If the movement is voluntary, then the signal results from a decision to do something – walk, run, jump, play the piano and so on. But the movement could also be an automatic response to some trigger – pain, tickle, losing balance – and so it may happen without our making a conscious decision.

In either case, the nerve signal is only the starting point for a carefully controlled sequence of events in the body involving nerves, muscles, joints, brain and inner ear. We are unaware of these separate events as we move our limbs smoothly and efficiently; without even thinking about it, we can touch, grasp, kick or push anything within reach with speed and accuracy, and usually with just the right amount of force required.

This fine control of movement is not with us from birth. A baby's first attempts at moving show a certain lack of finesse. He seems to know what he wants to do but does not know how to do it; he has not yet developed the methods to control his movements, although he has all the right physical parts.

It is easy to forget just how many moving parts we have – 112 in our wrists and hands, feet and ankles alone – when we are young. But some decrease in flexibility is inevitable as we get older.

Bones are joined to bones by ligaments, and with practice ligaments stretch, as demonstrated by the contortionist on the left. Is this the maximum flexibility the 26 bones of the spine are capable of?

Among the most remarkable joints in the body are those of the wrist, capable of the most intricate and finely controlled movements; the second joint of the thumb, a 'saddle' joint and an evolutionary breakthrough because it allows us to oppose our thumb and fingers; the rocking joint between the top vertebra and the skull, and the pivoting joint between the axis and atlas vertebrae, which give great freedom of movement to the head without damaging the delicate spinal cord as it enters the brain; the ball-and-socket type joints at hip and shoulder which allow movement in three planes.

The joints most vulnerable to wear and tear are obviously those that receive the most use – neck, hands, hips and knees.

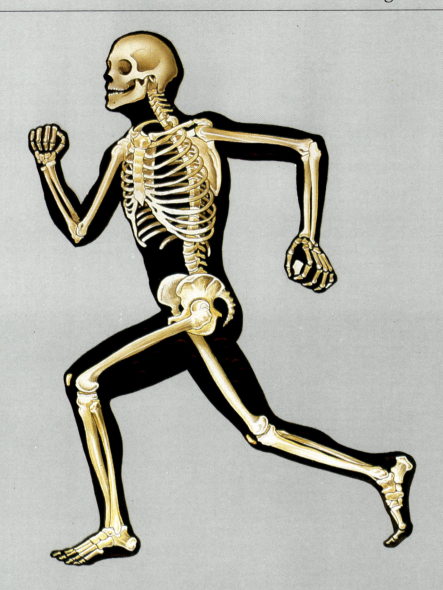

At the other end of life movements can also be difficult, but for different reasons. The older person has experienced years of fully controlled movements, with the benefit of an intimate collaboration between his nervous system and muscles. The problem now is that his joints have lost much of their smoothness and efficiency and he is unlikely to retain the ability to move his limbs through the wide range of angles that he used to be able to manage.

At its peak, when full control has been learnt and before the restrictions of ageing have set in, the fit human body can carry out a wide range of smooth and precise movements. Very few bodily movements are simple. To move a limb or part of the body to a specific point in space requires many different muscles pulling on many different bones. But to be successful, actions also require smooth movements about the points of contact between the bones – the joints.

The joints

There are more than 200 joints in the human body, each of them designed to allow more or less movement about a specific point. Some joints give a lot of freedom of movement – the shoulder, for example, is a ball-and-socket joint, the nearest there is to a wheel in the body. Other joints allow only a very restricted range. The elbow, for example, can only move in one plane, although it feels as if it can move in any direction. If you examine the elbow's movements more closely, you will realize that the shoulder joint plays a major part in movements of the lower arm. If you try to move your elbow without moving your shoulder you will see how limited the movement is. There are more than 100 joints in the vertebral column. Though the range of movement of individual vertebrae is small, except for those in the neck, their combined movements can dramatically arch or bend the back.

Although many of our joints are quite restricted in their ranges of movement, when they combine in an action they can provide almost limitless variety. A simple joint like the elbow is said to have one degree of freedom because it can move in only one plane; a joint such as the wrist has two, because it can move up and down, and side to side; and the shoulder has three because it can move in three different planes. This means that we can achieve complex movements – those involved in playing the piano, for example – by combining the separate simpler movements of individual body parts. Because shoulder, elbow, wrist and knuckles all contribute to a pianist's actions, if you look at the movements of the fingertips in relation to the shoulder, there is little restriction on how they move. By adding up the degrees of movement, or freedom, of the individual joints, it turns out that the fingers have 18 degrees of freedom, and we can use that figure to compare the versatility of a pianist's actions with simpler and more restricted activities.

Joint structure

Although the joints may seem simple linkages, they have a wide variety of complex structures. In man-made devices, some of the actions that a pianist makes can be achieved quite simply. Rods linked with simple pins, and rubber pads to protect the surfaces in contact, are enough to transmit the force of an action along a chain of rigid segments. But rubber perishes, metal corrodes, and friction erodes away points of contact. In the human body, every linkage point has to last for 70 years or more. Each joint has to be able to adapt to the growing size and strength of the body and the loads it imposes. To reduce wear and tear, the friction between surfaces must be kept to a minimum.

A joint like the knee has a structure that has evolved to overcome all of these problems. The two bones that meet at the hinge are held together by sheets of collagen, a tough elastic fibre. Inside the capsule of collagen, the bone ends are cushioned in a closed pocket of lubricating fluid. When the knee bends, the two bones do not come into contact, but are kept apart by the pressure generated in the fluid. The extreme slipperiness of this synovial fluid, as it is called, makes it as good as most man-made lubricating oils. In the joint, however, the oil never has to be drained and replaced, because it is continually recycled. Like most body fluids, the synovial fluid is filtered out of the blood plasma and has some special thickening ingredients added by the cells in the membrane that lines the joint. With joints like the hip, the knee and the ankle all benefiting from this kind of lubrication and suspension,

the body can withstand strenuous activity without jarring and damaging the joints. Some of these joints can take an impact strain of up to 20 times the force of gravity without any damage.

For 70 years or so this lubrication and suspension system generally works quite well. As we get older, however, movement of the joints can become less smooth and increasingly difficult. The inflammation that we call arthritis may occur in some of the joints, producing pain and sometimes damaging the ends of the bones. If there is also a reduction in the synovial fluid, every movement of the affected joint becomes difficult as bone rubs against bone. This is often the reason for the jerkiness and difficulty of movement in old people.

With a toddler, of course, this problem does not normally arise; his joints should work perfectly. Indeed, he can often do things that he will not be able to do when he is a heathy adult. Furthermore, he is born with some pre-programmed nerve–muscle control that is surprisingly sophisticated. For example, a newborn baby can make walking movements with his legs, even though it will be another twelve months before he can walk unaided. And he can also imitate an adult's facial expression – putting out his tongue, for example – which shows that there is some complex internal wiring in his nervous system that links the perception of someone else's tongue with what it feels like to move his own tongue.

Even when we are older, walking and other complex activities build upon innate abilities to perform basic movements and link up various muscle groups in different combinations. Although these actions are simple to describe from the outside, they involve complex sequences of internal bodily activities. Each of them involves thousands of cells and groups of cells and their associated nerve circuits.

Most movements do not involve only contraction of the muscles. It is usually the case that the contraction of one muscle to move a part of the body means that

The two circular diagrams above show the knee joint, protected by ligaments (red), which also loosely hold the kneecap in place, and enclosed in a capsule of collagen (pink). The bearing surfaces of the thigh bone and lower leg bones are made of cartilage, which is almost friction-free, but they do not actually touch. They are kept apart by the slippery fluid that fills the collagen capsule.

The two circular pictures on the right are of a damaged knee joint, seen from the inside, and of two knees mapped thermographically. The knee on the right is arthritic, and the hot, inflamed lining of the joint appears angry red and white.

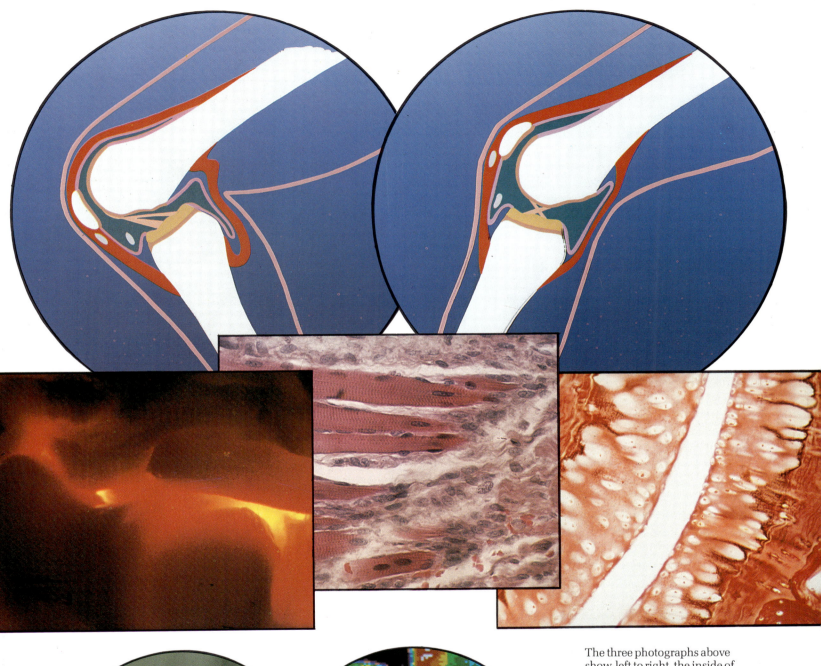

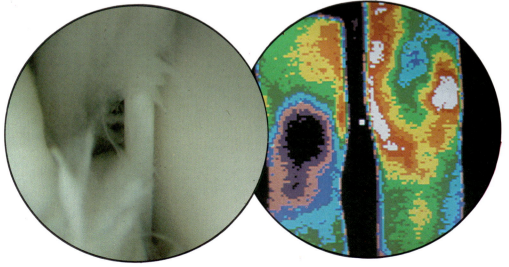

The three photographs above show, left to right, the inside of the elbow joint (the heads of the two forearm bones are at the bottom of the picture); a muscle-tendon junction (the long cross-striped bodies are muscle cells, and the fluffy-looking ones are tendon cells); and a finger joint, with the cartilage lining showing up well on either side of the joint space.

another muscle must relax. For example, bending the elbow by contracting the biceps muscle at the front of the upper arm means that the muscle at the back (the triceps) must relax to allow the arm to be pulled up.

Learning to move

As the growing child explores the world, he discovers the usefulness or the enjoyment of performing new actions and then repeating them. Since the knowledge of how to perform all movements cannot be built in at birth (how could every possible movement be anticipated by our genes?), the body clearly has the ability to learn and store complicated sequences of nerve messages and muscle contractions so that they can be put into effect whenever they are needed.

Even so, there has to be scope for variation. We never take the same step twice. The way we walk, for example, depends on many different factors. We have to be able to make continuous adjustments to the situation we find ourselves in. The muscles we use and the strength of their contractions depend on the slope of the ground, whether it is rough or smooth, soft or hard, whether we are carrying something, and so on. The brain has to take this sort of information into account when it issues instructions to the limbs to move. There must be two-way communication – it is not enough merely to be able to apply a force and hope that it achieves our intentions.

If we analyse a highly skilful human activity such as water-skiing, we can see how many different elements are involved in maintaining close control in the face of all the random variations of wind and water. First of all there is the need to start off smoothly as the towing boat speeds up and tension increases in the rope. With the ever-changing variation in this tension, the skier must be able to know how much pull to exert with his muscles, so that it is just enough to counteract it. If he is not aware of the moment-to-moment changes in tension as the boat's speed and the water texture vary, he will be unable to adjust his own pull accurately enough to keep his balance. He needs feedback of information from his muscles to his brain.

For this feedback to work, the first piece of essential information is the location of the various parts of the body. Although the skier can see where his hands and legs are, he cannot spend all his time looking at them – his vision is more important for other purposes. In fact, we are aware of the position of our limbs and joints at every waking moment, although we do not always use the information. If you try to touch your nose with your eyes shut, for instance, you will find you can do so irrespective of the starting positions of your hand and head. You can manage this feat because there is a

Balance and coordination are achieved by synchronizing countless nerve signals with the movements of hundreds of different muscles, and almost instantaneously feeding movement information back to the brain so that minute corrections can be made.

continual stream of messages as the hand moves that tell you where it is at any one time.

Every muscle has inbuilt receptors that continually feed back messages to report on the effect that the muscles are having. Some receptors tell how long a muscle is; others tell how taut it is; and yet another type of receptor feeds back information only when the length is changing. The combination of signals from these receptors can help the brain to determine where the limb is and can also help it to accurately judge the amount of muscle tension needed. If a force is exerted too strongly or too quickly, we usually notice its effects so quickly that we have time to modify our actions.

The tension receptors near the point at which the muscle is attached to the bone have an extra function: they can warn when tension is so great that it might break the bone or pull the tendon apart. Sometimes weightlifters manage to anaesthetize these receptors in order to get that extra bit of muscle power. Unfortunately, by silencing the inbuilt alarm system, they sometimes end up with broken bones.

It seems, then, that with these two mechanisms – feedback as to where the limbs are, and linked sequences of muscle actions – we can adequately explain how smooth and appropriate movements occur. Unfortunately, these two factors are not adequate to explain very fast activities, finger movements of an expert musician, say. With many different muscle movements every second, there is too little time for feedback from one movement to reach the brain so that the next movement can be modified accordingly; nerve messages travel too slowly. If playing a musical instrument could only be carried out on that basis, we would never hear any music faster than a slow march.

A musician starts by learning each movement separately and slowly, relying on feedback from his muscles and joints to tell him whether he is getting it right or wrong. This is inevitably a slow process at first, because it takes time for nerve impulses to travel between the limbs and the brain. But with practice the musician plays faster and makes fewer mistakes; this happens because the brain does not need such detailed feedback – it has learnt the correct muscle and joint actions. We acquire skills like this by a slow and arduous process of learning.

A virtuoso performance of Liszt's paino study *La Campanella* probably represents the utmost coordination that the human hand, eye, ear and memory are capable of. The octave leaps and repeated notes in the right hand are performed at terrific speed.

Balance and movement

So the feedback that helps us to learn and carry out body movements contains information about position, speed and even acceleration of the various parts of the body. But there is another sense that is essential to carry out coordinated actions – the sense of balance.

The body needs to have some way of knowing up from down. It also needs to know about the angle of the head and its direction of movement, so that it can be kept upright whatever the rest of the body is doing. In the inner ear are a number of small chambers lined with a 'carpet' of hair cells embedded in gelatine. On top of this carpet lie crystals of calcium carbonate which shift about as the head moves. These movements are transmitted through the carpet of hair cells to receptor cells which send nerve signals to the brain.

While these devices tell the brain about the angle of the head, there are others that specifically detect acceleration. These take the form of three bony hoops (the semi-circular canals) in each ear. These are filled with a watery fluid which moves when the head rotates or undergoes acceleration. Movement of the fluid stimulates receptor cells in flaps of tissue in the semi-circular canals, sending nerve messages to the brain about the direction and magnitude of the head's movements.

The way this system works can be illustrated by imagining what happens to a glass of water in the centre of a turntable. As the turntable starts to rotate, the liquid stays still and the glass rotates around it. Then the liquid begins to rotate as well, eventually catching up with the glass. And when the turntable and glass stop rotating, the liquid continues to rotate for a while. The fluid in the semi-circular canals behaves in a similar way, the receptors in the flaps of tissue in the canals responding to the speeding up and slowing down of the fluid.

As the head starts rotating, the fluid lags behind and bends the flap one way, triggering nerve messages to the brain. Then the fluid catches up with the rotation, and the flap moves back to its normal position. Finally, when the head stops rotating, the fluid continues to move and bends the flap the other way.

When someone spins very fast, his eyes flicker from side to side after he has stopped. This happens because the fluid in the ears is still moving and still sending movement messages to the brain which, in turn, is causing the eyes to flick backwards and forwards over a world that is no longer flashing by. These signals also produce the unpleasant feelings of nausea that are part of motion sickness. The way an experienced dancer or skater overcomes this irritating problem is by moving the head so quickly and suddenly on each rotation that

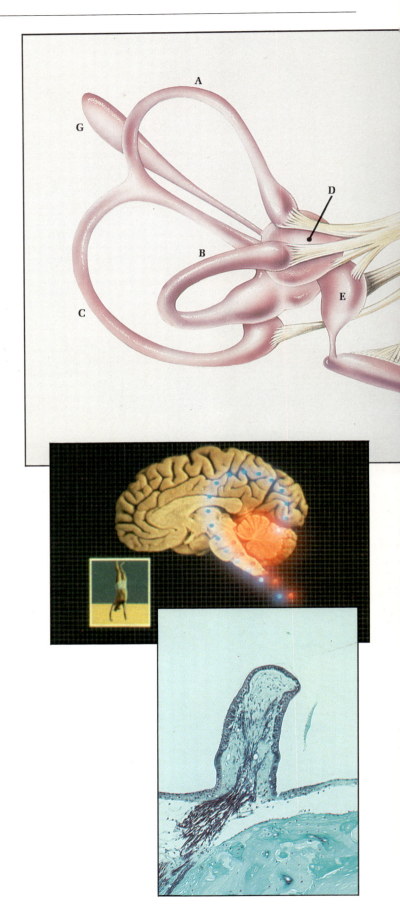

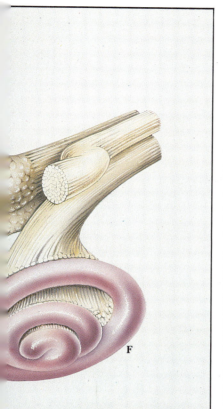

The organs of balance, the three semi-circular canals **A**, **B** and **C**, and the utricle and saccule **D** and **E**, are shown on the left. They are associated with the cochlea **F** in that they sit in the same bony labyrinth inside the skull and are filled with the same fluid, called endolymph. This fluid is produced by special cells lining the organs of balance and hearing and is removed by other cells in the blind-ended endolymphatic duct **G**. Note the nerves supplying the bulbous ends of the three semi-circular canals and the utricle and saccule.

It is in the cerebellum, shown below right in cross section and left in relation to the rest of the brain, that coordinates information from the organs of balance with information from the other sensory organs and from the rest of the body. Pressure receptors in the palms of our hands also feed balance information to the cerebellum and help us to stay upright in a handstand.

the liquid does not have time to move and so no signals are sent to the brain and the whirler does not become hopelessly giddy

The brain's role

The more that is known about what is involved in even the simplest movement, the more surprising it is that any of us can move at all. The computing tasks that are required, moment by moment, to keep us upright and active require a large amount of brain tissue. And in fact there is a separate section of the brain, the cerebellum, devoted almost entirely to controlling movement.

The cerebellum, whose name means little brain, is an outcrop on the brainstem. Situated at the base of the rear of the brain, beneath the bulge of the cerebral cortex, it monitors the nervous activity that is going up and down between the brain and the muscles. Although smaller than the two cerebral hemispheres, the cerebellum actually contains far more nerve cells than they do, reflecting the complexity of its tasks.

The cerebellum is unable to initiate any movements, but it is responsible for the smooth running of all our complex muscular activities, and it achieves this by changing the instructions that the cerebral cortex sends to the muscles.

Research shows that the cerebellum receives a constant stream of incoming sensory information that enables it to monitor the position of various parts of the body. It also receives copies of information from the cortex about the movement commands it is issuing to the muscles. Using these two sorts of information, the cerebellum can predict the movements and positions of different parts of the body, and then decide whether the movement being produced is the one that is intended. If not, it can actually correct the movement by intercepting the messages from the cortex as they travel down to the muscles.

So however simple a bodily movement may seem, its execution usually depends on many elements for its success. Cortex, cerebellum, nerve cells, muscle receptors, muscles, organs of balance, bones and joints all play a crucial part in the chain of bodily events that leads from decision to execution of a human action.

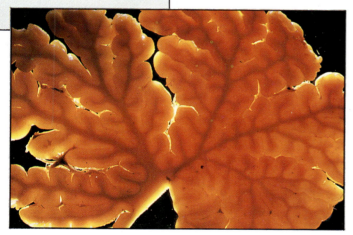

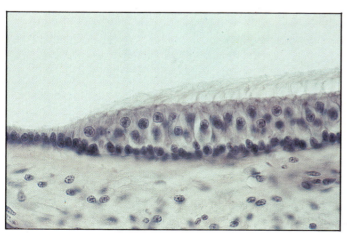

On the far left is a view of the tiny crest of tissue in a semi-circular canal; the hair cells on its surface detect the slightest acceleration or deceleration in the endolymph that bathes them. On the left is a portion of the hair cell 'carpet' that lines part of the utricle and saccule.

13 · The Depths of Sleep

What is the living body *doing* when it is asleep? Apart from an occasional sigh, snore or grunt, it makes no noise, and only the sporadic shift of the body or turn of the head shows that it is capable of any movement. The owner of the body appears to be playing no part in the world he lives in; in fact he is oblivious of it. And he and the rest of us are in this state for one third of our lives – more time than we spend on eating or reproducing or other activities that are apparently more important for our survival.

Sleep is obviously an essential part of life. No one can do without it, and the urge to sleep is so insistent that no one can resist it. We can voluntarily starve ourselves or deprive ourselves of drink until we die, but we cannot force our bodies to go without sleep indefinitely.

So what is it that makes sleep so important ? In this chapter we shall look at the strange combination of events that happens in the body when we go to sleep, and at how successful the body is at looking after itself, when our minds are far away in the land of dreams.

Sleep takes place at various levels. We recognize this when we talk about being 'deep asleep'. There is sometimes a sense of struggling up to the surface when we are suddenly woken up, and there are other times when we feel just on the borders between sleeping and waking. These feelings we all have correspond to the findings of scientific research on brain activity during long periods of sleep. Scientists now recognize several levels of sleep, which are marked by different electrical activity in the brain and by different events in the body. In fact, the more closely we look at sleep the more we find going on beneath the surface.

Stages of sleep

A night's sleep is like a voyage, with a beginning, a cyclical middle and an end. While our sleeping brains

go through a series of well-defined stages of activity, our sleeping bodies also follow a predictable pattern, in which hormones, nerves, digestion and growth all have a part to play.

Some clues to what is happening when we sleep come from the electrical activity of the brain. Every nerve cell produces detectable electrical voltages whenever it is active, and scientists are sometimes able to measure this activity in single cells, although they have to penetrate beneath the scalp to do so. The combined activity of many cells is easier to detect, because the total electrical activity of the brain can be measured through the thickness of the skull. A signal detected in this way seems random because it is made up of separate signals from millions of nerve cells all doing different things.

The combined signal is like the noise from a room full of people all holding different conversations. At first, the variations in the sound have little meaning for a listener, although the general pitch of the signal might tell you whether there were more men than women in the room, for example. Occasionally there might be larger variations in volume. Perhaps someone particularly

interesting or famous comes into the room and the sound drops for a moment before continuing. The situation starts to be really interesting when the people in the room make similar noises at the same time. If there is cause for celebration, they may all sing the same song, at the same time, and the pattern of sound will be noticeably different from the general babble that was detected before.

Analysis of these brain waves shows a distinct pattern of brain activity that begins even before we fall asleep and continues through the night in a series of ups and downs until we surface again eight hours or so later.

The first thing that happens, as we settle down and begin to feel drowsy, is that the random pattern of brain waves that occurs when we are active is replaced by a more regular pattern as some of the brain cells start to act in synchrony. This is called alpha rhythm, about 10 cycles a second, and is easily disrupted by outside noises or by the need for some active thinking. During this alpha activity of drowsiness, rapid waves pass across the brain showing the switching off of activity in those sections of the brain connected with attention.

After a while, we *suddenly* pass from waking to

Over the last 30 years sleep laboratories have produced a lot of information about what goes on in our bodies during sleep, most of it based on measurements of the electrical activity of the brain. Volunteers have electrodes attached to their scalp before they go to sleep. Even through the bony thickness of the skull the electrodes are sensitive enough to detect variations in activity in various areas of the cortex. An EEG (electroencephalograph) machine amplifies these rhythmical electrical signals up to two million times and translates them into pen-tracer waves on graph paper. Nevertheless it takes a skilled analyst to pick out the predominating frequencies. These change very markedly during the course of a night's sleep.

EEG recordings have cast some light on sleep disorders such as rocking, groaning, sleep-walking, nightmares and narcolepsy. The researcher in the picture above is attaching electrodes to a man who suffers from narcolepsy, a disorder which causes him to fall asleep at odd times and for long periods, especially after feeling strong emotions.

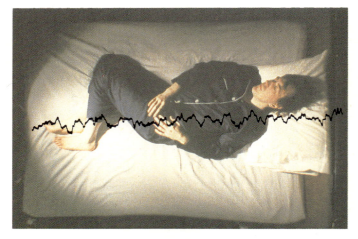

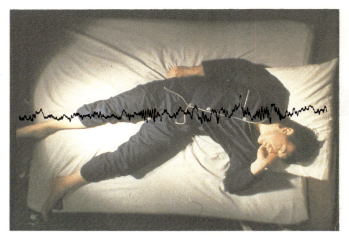

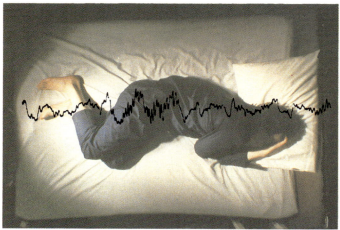

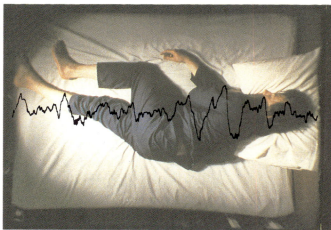

It takes about 90 minutes for us to reach the deepest stage of sleep. In the first few minutes (top left) our brain waves show some of the small irregularities typical of the waking brain. As we fall deeper into sleep (top right) the waves start to become bigger and slower, punctuated by bursts of spindle-like waves. Deeper still, above left

and right, the fluctuations in the wave trace become even larger and slower. In the deepest stage of sleep, heart rate and breathing are slow and steady and our muscles are completely relaxed. Throughout the night we oscillate between deep and shallower levels of sleep, with increasingly long periods of dreaming sleep in between.

sleeping. There is no period of transition – one moment we are awake, the next we are asleep, although (as many children who have tried find) the actual moment of falling asleep is impossible to detect.

Once we are asleep, we experience a series of different types of brain activity that produce bigger and slower brain waves than we ever have when we are awake. Initially, the electrical activity produces waves at a steady, slower rate. As sleep becomes deeper and deeper, the whole brain seems to throb with a gentle and unspecific type of wave activity, at a frequency of one or two cycles a second, which suggests that most of the cortex at least (the surface of the brain), which is where the waves are measured from, is reducing its activity. This stage lasts about half an hour or so.

Then, very quickly, a whole series of events takes place. The brain waves start to become more active and produce waves of a higher frequency. Instead of many

brain cells doing the same thing, they have now split up into much smaller groups and are as active as when we are awake. If you were watching the sleeper, you would see what appeared to be signs of wakefulness. The heart rate goes up, there is a slight increase in the rate of breathing, and the eyes move rapidly under the eyelids. In both men and women there is increased activity of the sex organs, with erection and secretion.

This is the most interesting phase of sleep, called rapid eye movement (REM) sleep. From now on, throughout the rest of the night, the sleeper will alternate between bouts of REM sleep lasting 20 to 30 minutes and hour-long periods of deeper sleep. It seems to be an essential part of our sleeping pattern, and is observed in everyone. Surprisingly, these rapid eye movements were first noticed only 30 years ago. Insomniacs have had thousands of years to observe their sleeping partners, and yet no one suspected that the

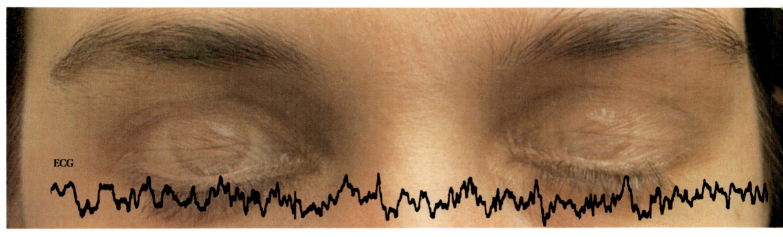

ECG

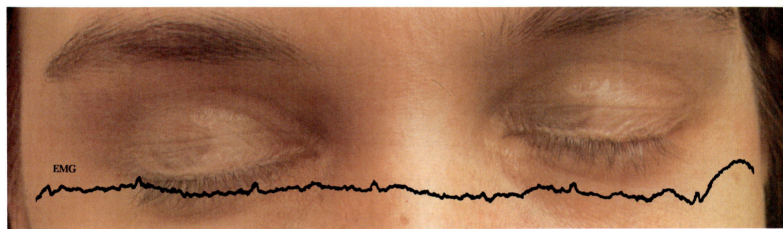

EMG

During REM sleep, which occurs every 90 minutes or so on average, the eyes twitch about under the eyelids as if following a fast-moving film. If a sleeper is woken during a bout of REM he or she usually reports a vivid dream. We all go through periods of REM sleep, even if we do not remember dreaming when we wake up in the morning.

Many animals show REM too. Periods of REM sleep get longer as the night progresses.

The traces above show the ECG and EMG during REM sleep. The EMG (electromyograph) trace shows an absence of muscle tone, because the brain stem blocks movement messages.

seemingly monotonous eight hours of sleep were broken up into such distinct episodes.

Dreams

There has, of course, always been one observation that we can all make about our sleeping activities – the occurrence of dreams. The periods of rapid eye movements seem to coincide with the periods when the sleeper is dreaming; people who are woken up while in REM sleep almost invariably report that they were in the middle of a dream, while people woken during non-REM sleep usually report no vivid mental activity. Disappointing though it may be, there is no scientific evidence that dreams provide a significantly different type of mental activity from that of the waking mind. Of course we are freed in dreams from inhibitions and legal constraints in the activities we dream of performing. But such fanciful ideas as precognition or telepathy owe

more to coincidence than to any demonstrable mental process. And even the undoubted property that dreams have of providing material for psychoanalysis is similar to the way our waking mental lives also provide a window into the unconscious. The correct explanation for the purpose of dreams may be a more prosaic one – perhaps the brain benefits from more unusual mental experiences than real life can provide, or perhaps, as some scientists think, dreams serve to sort through the mental events of the day, storing some elements and rejecting others.

Some scientists have even suggested that the direction of the eye movements has something to do with the content of the dream. When a group of sleepers was observed over a period, most of them produced mixtures of vertical and horizontal movements of their eyes but, in one group of experiments, several subjects showed purely up-and-down movements of the eyes

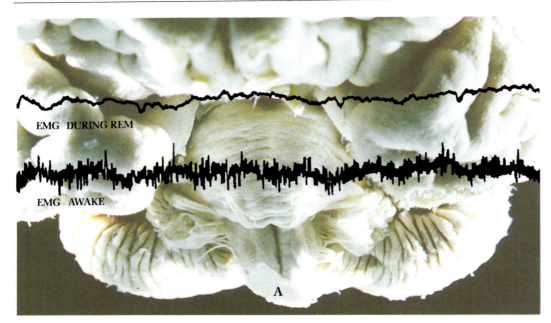

EMG DURING REM

EMG AWAKE

A

During REM sleep a remarkable change comes over the muscles of the body. Most of the time our muscles have some tension in them, but during REM sleep the brain stem damps down all muscle activity so that we are paralysed. This prevents us acting out our dreams. The brain stem **A** is shown poking down from the underside of the brain on the left, overlaid by two EMG traces, one showing awake muscle tone, the other REM muscle tone.

Yawning is a mysterious and infectious phenomenon that often precedes sleep. The deep breath that a yawn produces probably opens up all the alveoli or air sacs in the lungs before we settle down to a night of shallow breathing.

and when they were asked about the content of their dreams they had one interesting feature in common. One subject had been dreaming of standing at the bottom of a cliff and looking up at climbers; another had dreamt of climbing up a series of ladders; and a third had dreamt of throwing basket balls at a net and then looking down to pick up another ball from the floor. If these results mean what they seem to mean, they represent a kind of acting out in bed of the events of our dreams. But they are acted out on a very small scale.

During REM sleep, very little movement of the body takes place, apart from the eyes. Measurements of muscle activity show that all activity stops quite suddenly during REM sleep. There is a part of the brain stem concerned with movement which sends out a general message damping down all the main muscles of the body. One useful by-product of this is that any movement instructions our brains are sending out during dreams do not get beyond the brain stem to the muscles. This means that we are unable to act out even the most violently active dreams.

Some idea of how inconvenient it would be to be without this mechanism comes from experiments on cats, in which their normal dream paralysis was overcome surgically. Their brains showed all the signs of REM sleep but their bodies were active, performing normal daytime activities like chasing imaginary prey or recoiling from a threat. Watching these animals, it is difficult not to believe they are acting out their dreams.

There are of course occasions when activity does occur during sleep. But sleep-talking, sleep-walking and some dreams and nightmares actually occur in non-REM periods of sleep and the sleeper is much less likely to report vivid image-filled dreams. More often he talks about feelings of dread or being suffocated.

Body maintenance

This rich mental life we lead during our sleep is accompanied by various internal bodily activities. Some of these activities happen mainly during sleep, as if to use the special opportunities presented by the resting body. Others are part of the continual, life-long maintenance of the internal environment that must not be allowed to stop for a moment.

While we are asleep, the body steps up its activities in a number of areas relating to growth and development. For example, during the night, all of us produce larger amounts of a hormone called growth hormone, which controls the rate of growth of bones and also affects the rate of metabolism. Within a couple of hours of going to sleep, a sleeper will start to secrete high levels of growth hormone and this will continue on and off throughout the night. In addition, there is evidence of increased chemical activity in the brain, as it synthesizes new proteins that will be needed for its waking activities. This can take place at five to ten times the waking rate, almost as if the brain was seizing its chance to make use of a period of less external activity. Indeed, it may well be that the purpose of sleep is to allow activities involving synthesis and repair such as these to be carried out free from the disturbances of waking life.

There are also many complex activities which must go on all the time, day and night, and which need a sophisticated control system. The heart has to keep beating, oxygen must continue to circulate, carbon dioxide must be disposed of. Temperature must be

controlled, the acidity of the blood must not vary much, and even such tasks as digestion must continue, if we are not to find our supper still in our stomachs at breakfast time.

All these maintenance tasks are supervised by the autonomic nervous system which links the lower parts of the brain and the spinal cord with many parts of the body that have to be able to work on their own without conscious control. If we had to plan each breath and heartbeat and then decide to carry them out, we would have little brain-power left to do anything else. If we had to guide food through our alimentary canals by consciously squeezing it along every few minutes, and control the rate at which urine flowed out of the kidney, we would have little time left to enjoy life. Instead, each important body system has a dual nervous supply, almost like a brake and an accelerator that includes one particular segment of the nervous system. This is usually a section of the spinal cord, capable of monitoring and adjusting the activity of its particular organ or system without needing to refer upwards to the brain.

Of course, many of these tasks have to be carried out in the daytime as well. What we see at night is the base-line, the level at which the body has to tick over in order to keep going at all. When we are awake, the same careful automatic control of activities has to continue, but it is often complicated by changes which are necessary for us to be able to act – increases in heart rate for exercise, changes in digestive activity as we eat or drink, and so on. And when we *are* asleep, these activities interfere as little as possible with a healthy night's sleep.

Waking up

As the night wears on, the periods of deep sleep become shorter. It is as if we are gradually drifting nearer to the surface. Somehow or other, the body knows when it has had enough sleep. For some of us that could be after five or six hours; for others nine or ten do not seem to be enough. Sooner or later, however, we all wake up.

Even before we wake up, there is a part of the brain that monitors sensory messages from the outside, and maintains the vigilance that may be necessary for our survival, or someone else's. The most deeply sleeping mother, for instance, will awake to the distant sound of her baby crying.

All night the body has been receiving a mixture of sensory messages, ranging from the pressures of the bed and the temperature of the air to night-time noises. Sometimes these messages can penetrate our consciousness quite easily – the part of the brain dealing with hearing is almost as sensitive during REM sleep as it is when we are awake. But that does not necessarily mean that we will note the content of those messages. Only a few will be important enough for the body to have to wake up.

To separate out the few important messages, a network of cells in the brain stem monitors all incoming sensory signals. Almost like a gatekeeper, it prevents the normal and unsurprising messages from going any further and disturbing the sleeping brain. Presumably, it can compare each message on the basis of previous experience. After all, someone who lives near a railway line will sleep through the loudest of trains, while a guest will be woken up all through the night by the same noises. Similarly, the baby's cry will be out of the ordinary and signify a need for urgent action.

There is still much that is mysterious about sleep. Even its physical causes are unknown. Some research suggests that we sleep because of the build-up of a chemical in the blood, and yet Siamese twins, who have the same circulation, feel sleepy at different times.

The purpose of sleep and dreams is also shrouded in mystery. Sleep *may* be for the purposes of rest and repair, but not all animals sleep – sharks do not, for example. It seems to be clear, however, that we have a use for both sleep and dreaming. People who say they do not dream nevertheless go through periods of REM sleep and can sometimes be persuaded that they have dreamt if they are woken up in mid-cycle; and people who are deprived of REM sleep seem to need to catch up the following night by spending a larger proportion of their time in that state. Similarly, those who say they never sleep have been shown to take short, frequent naps without realizing it.

14 · Hot and Cold

We live in a world of extremes of temperature, from arctic cold to tropical heat. But in spite of the large differences in the temperature of the environment, man's body temperature is remarkably stable. And he maintains that temperature wherever he goes in the world. Of course, some of this is achieved by using our intelligence – to make clothes or build shelters – but there is much the body can do for itself to keep cool or stay warm in the face of a changing environment.

The human body temperature is set at the best level for its chemical processes to work efficiently. It is kept that way automatically by the interplay of many different systems in the body. Such familiar activities as sweating and shivering only make sense when we see them as parts of our well-organized temperature-control system.

The natural tendency of heat is to flow to colder objects. Abandoned in a cold world, the warm human body would soon lose all its heat, and that would be disastrous for the internal environment of its cells and tissues. If our surroundings are too hot, we are also faced with a problem – how to prevent the build-up of excessive heat in the body. But the human body is so versatile that in the course of a day's skiing, for instance, it can speed from icy mountain top to sauna bath and still maintain a core temperature that hardly varies at all.

Heat creation and loss
The body preserves its stable temperature by maintaining a balance between two processes: creating heat and losing it.

To create heat, the body fuels its heating system with

It seems that our earliest ancestors were dark-skinned and heat-loving. The challenge of colonizing temperate and subarctic regions was to find ways of conserving body heat.

food. Most of the food we eat is turned into heat and only five per cent is used to perform useful work. After it has been digested in the intestines, the component molecules are delivered to various organs and systems, where they are burnt and produce heat.

An organ like the liver, for example, is a hotbed of chemical activity: in the space of an hour it gives off as much heat as a one-bar electric fire gives off in a couple of minutes. Because the liver is well supplied with blood vessels, the blood that passes through it can carry the heat away. The liver is like a boiler in a central heating system, and its heat is distributed to the body's 'radiators', the skin.

Another major source of heat is the muscles. Whenever we do anything energetic, they burn up glucose to produce the energy required. That burning produces heat, which is carried away in the blood.

Even tiny organs such as the endocrine glands add a significant amount to the body's heat production, as their cells work hard to secrete hormones – our chemical messengers.

All this heat is used to keep the body's core a steady temperature. Here, in the central area of the trunk, are most of the important organs, which work best in a stable environment of 37°C (98.4°F). (The brain is the only major organ outside the core, and that has its own local temperature regulation system.) Outside this area are layers of flesh and less important body systems, which act as insulation, protecting the boiler at the centre of the body. Minor temperature fluctuations do occur during the course of a day. (For example, we are cooler when we wake up than we are later in the day; this could be because our daytime activity generates a certain amount of surplus heat.) But nonetheless our core temperature is kept remarkably steady – to within half a degree or so.

Further away from the core of the body, in the limbs and layers of fat, the temperature varies much more. The temperature is not so crucial here because there are no vital organs to be affected by temperature changes. The coolest part of the body is the feet. They can be at least 10°C (20°F) cooler than the normal core body temperature without coming to any harm. The hands can also be cooler, but tend to stay warmer than the feet.

The hypothalamus in the brain (located near the white oval patch at the top of the spine in the scan above) ensures that core body temperature and brain temperature remain steady.

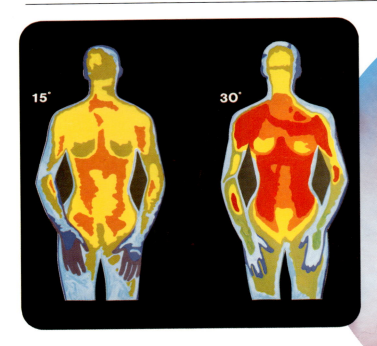

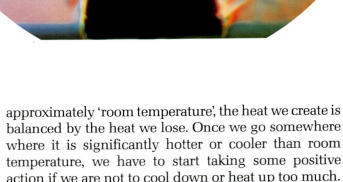

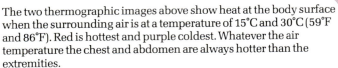

The two thermographic images above show heat at the body surface when the surrounding air is at a temperature of 15°C and 30°C (59°F and 86°F). Red is hottest and purple coldest. Whatever the air temperature the chest and abdomen are always hotter than the extremities.

These variations depend on the temperature of the environment. A normal room temperature is about 20 degrees cooler than the core temperature of the body, and so we are continually losing heat to our environment in several ways.

When we breathe in air that is cooler than body temperature, the body uses heat to warm it up inside; when we eat cold food or drinks, these take in heat as they travel down the gullet; when we breathe out, we are giving up some of our body heat in the expired air; and faeces and urine take with them some of our body's heat production.

A naked body will lose heat by radiating it to any nearby cooler object. It will also lose heat by convection – any air that comes in contact with the body will heat up, rise because it is warmer, and allow cooler air to come in and take away more heat. Interestingly, because warm air rises, we lose more heat by this process if we are lying down than if we are standing up. This is because the air currents carry heat away from every part of the body simultaneously when we are lying down, whereas when we are standing up the heated air from lower down travels up over the warm flesh above and protects it from rapid heat loss.

When the temperature of our environment is approximately 'room temperature', the heat we create is balanced by the heat we lose. Once we go somewhere where it is significantly hotter or cooler than room temperature, we have to start taking some positive action if we are not to cool down or heat up too much.

Coping with a cold environment

The most effective ways we have of coping with temperature change are, of course, actions we take ourselves. We would be foolish to get out of our beds and go straight outside into a snowy environment expecting the naked body to adapt on its own. Most of us, particularly if we live in climates with a wide variation in temperature, can choose from a range of clothing. By adapting his clothing, man has managed to survive in temperatures as cold as −60°C (−76°F), in the Arctic, or as hot as 200°C (392°F), on the surface of the moon.

The purpose of clothing is to prevent or slow down heat loss by radiation or convection (or heat gain if we were on the moon). As well as preventing cold air from coming into contact with the skin, clothes trap a layer of warm air which helps to insulate us. In using this method of keeping warm, we are imitating the coats of animals, which function in a similar fashion.

In addition to wearing clothes, we can protect our

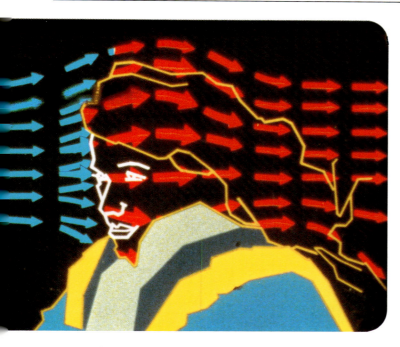

Heat can be carried away from a hot object, such as the boiling kettle on the left, by the air that surrounds it. This process is called convection, and it can have a marked effect on body temperature. Normally a layer of warm air clings to the skin, but if it is replaced by cold air, as happens when it is windy, skin temperature is maintained by producing more heat inside the body, an effect known as the wind chill factor.

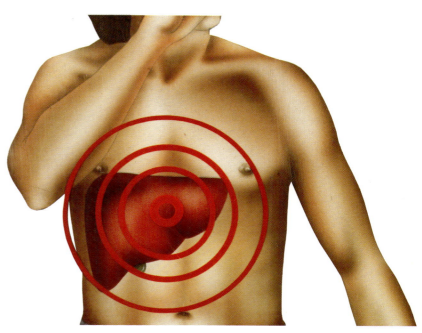

We all have a source of central heating in the core of our body. The liver, and to a lesser extent the other abdominal organs, generate heat as a by-product of the breakdown of food. At rest, the abdominal organs are responsible for more than half the body's total heat output. They also help to make good moderate heat losses.

bodies from heat loss by various types of behaviour – by lighting a fire, moving to a warmer place, or curling up in a ball to reduce the surface area from which heat might escape.

For us to do this, we have a thermostat that is set to a certain level, and we have sensors in the body that tell us if we are in danger of deviating from that level. Each of us has a personal thermostat at the base of the brain, in a tiny but important section of nervous tissue, the hypothalamus.

Since it is most important that the temperature of the trunk is kept constant, the hypothalamus compares the actual core temperature with what it should be. There are cells in the hypothalamus that are bathed in blood flowing from the innermost parts of the body and can sense any change in the blood temperature. There is also another kind of sensor in the skin that provides back-up for the hypothalamus and plays a particularly important part in the body's response to cold. These are the temperature receptors, which respond to changes in temperature by firing at different rates along nerve fibres that end up in the hypothalamus. One type fires rapidly when it is below body temperature and less rapidly as it gets warmer; the other fires more rapidly in the warm and slows down as the temperature drops. If

the environment becomes cooler, signals will be sent from the first type of receptor in those areas of skin that are exposed.

Although there is no immediate change in blood temperature, a number of alterations will be set in motion. A signal goes out to constrict the blood vessels in the skin, so that less warm blood reaches the surface. This is why the skin can go pale in the cold. At the same time, the tiny muscles that surround the hair follicles make the hair stand on end, producing 'goose-flesh'. This is a feeble remnant of our ancestors' ability to raise their covering of body hair to trap a greater layer of air to improve their insulation.

If our surroundings are both cold and windy, the body loses heat even more quickly because the protective layer of air around the body (which has been warmed by heat radiated from the body) is blown away by the wind. As the temperature drops further, more drastic measures are needed. As well as conserving heat, the body must attempt to create more heat.

One way in which the body achieves this is by starting a rather uncoordinated series of involuntary muscle contractions throughout the voluntary muscles. Because they are uncoordinated, they do not result in any major movement of an arm or leg, but in a series of little contractions which we know as shivering. The heat released by these thousands of muscle fibres contracting is carried away by the blood to raise the general temperature of the body. Almost every muscle in the body can shiver – the only exception being the eyes.

Another method of creating heat is by exercise. As we have seen, muscle activity creates heat, and this is why we find ourselves indulging in apparently pointless stamping and hugging movements on cold days. Our voluntary muscles can increase their heat production tenfold in situations like this.

The body also has internal devices that can help to avoid heat wastage. The flow of blood, which distributes heat, is from the heart outwards, through the arteries, and from the surface inwards, through the veins. At low environmental temperatures there are quite big differences (up to 12°C or 22°F) between the core and the surface and even between different parts of the surface. Without regulation, hot blood would continue to flow to cooler areas such as the limbs, where the heat was not so vital, and cool blood would flow inwards to areas that need to retain heat. Fortunately, there are parts of the circulation where arteries and veins run side by side, almost intertwined, with the flow of blood in opposite directions. This means that the arterial blood can warm the venous blood by heat exchange, and heat that would have gone on to the surface can be returned to the core of the body where it is really needed. To help this process, the arteries can constrict to slow the blood down and give it more time to give up its heat.

The heat that is distributed in this way is produced in the body at a certain rate. If the cold conditions continue for long enough, a hormone is released which produces a general increase in this rate all over the body, in all the tissues. For as long as it is cold outside, these very efficient heat-producing and heat-conserving measures will continue.

The stimulus for all these events is the arrival of cooler blood in the hypothalamus or the superficial sensations of cold on the surface of the skin which trigger messages to the hypothalamus. And the sensation of cold on the skin is unpleasant, so that we also take some voluntary measures of our own, rather than leaving it all to the automatic systems of the body.

Combating overheating

The effect of a warm environment, in contrast, is often to produce a pleasant sensation on the skin, and not usually something we feel an immediate need to correct voluntarily. However, the hypothalamus detects the slightest rise in the temperature of the blood and sets in motion a whole chain of actions to reduce the heat as fast as possible.

As with the response to cold, the first signs of this action appear in the skin. This time, the blood vessels in the skin dilate, so that more of the body's heat can be radiated away. The redness that comes over our skin when we are hot is an indication of the increased blood flow, and indeed the skin itself feels hot to the touch, not just because it is hot outside but because the body's core heat is radiating out through it. We can also increase the amount of heat lost by convection by exposing more skin to the air. Radiation and convection, account for about three-quarters of our total heat loss.

The rest of our heat loss takes place through evaporation. To change water from a liquid to a gas uses energy, or heat, so the heat that is brought to the surface in the blood vessels of the skin goes into evaporating the water that comes from our sweat glands, instead of raising the body temperature.

The skin is giving off water vapour all the time, even when we are not in a particularly hot environment – it is just one way of making sure that heat does not build up inside. But when the heat increases, more water has to be poured onto the skin surface to increase the rate of heat loss by evaporation. We call this sweating.

Sweat glands are spread over the entire skin surface, and are particularly concentrated on the forehead, exterior face of the thighs, palms and the soles of the feet. The total weight of the sweat glands of the body is 100 grams ($3\frac{1}{2}$oz) – two-thirds that of one kidney. There are actually two types of sweat glands: the ones all over the body, which help us to keep cool; and others concentrated in the armpits and around the groin, which secrete another, non-cooling kind of sweat, containing a number of different molecules, including the distinctive personal scent that every individual has.

The cooling sweat glands produce a solution of salt and water which travels along a tube to the surface of the skin. The fact that it has to travel some way to reach the surface enables the liquid to lose some of its salt, which is reabsorbed into the body. If the sweat was excreted straight from the gland to the skin surface we would lose a lot of salt along with the water with a resulting decrease in salt concentration in the blood.

Sweat also carries a substance that helps to open up the blood vessels even further so that more heat-carrying blood reaches the surface. This means that the heat exchange between arteries and veins that saved heat in the cold is minimized, so that cool blood goes to the core and hot blood continues to the surface.

One third of the heat lost by evaporation leaves the body through the lungs. Even when you cannot see

When it is cold, bare skin loses a lot of heat. A bald head is particularly wasteful of precious heat, as this Schlieren photograph shows. Because heat rises, convection is strongest at the apex of the head. A hat, like other clothes, cuts down heat loss because it traps warm air next to the skin.

We also lose heat every time we exhale. The gentleman below is turning this fact to good account by blowing on his cold hands.

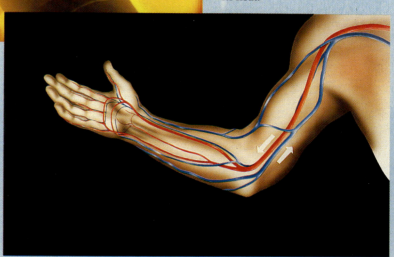

To keep blood flowing to and from the heart and brain at an even temperature, we exchange heat between the arteries and veins of our limbs and extremities. The diagram below illustrates the principle: blood flowing away from the warm core of the body gives up its heat to cooler blood returning from chilly surfaces. In cold weather our arteries make this process more efficient by constricting; this slows down the blood so that it has more time to give up its heat.

Working muscles generate heat as a by-product of burning glucose. During strenuous exercise in cold weather our muscles generate up to 90 per cent of our total body heat.

Body temperature can also be affected by food and drink. The cup of coffee in the picture on the right shows up white, and the parts of the body warmed by it – lips, teeth, throat – show up red or orange.

If we did not have sweat glands, shown in cross-section below the cup of coffee, we would be reduced to panting like dogs when we overheated. We have sweat glands all over our body, but those in the armpits and groin are larger and produce the musky odours we try hard to mask with deodorants. At one time these secretions may have been important in courtship and mating.

As sweat evaporates from the skin (above), it uses up heat energy and cools us down. This mechanism is so efficient that we can temporarily withstand temperatures above the boiling point of water, as in a sauna.

When we are hot, our blood vessels dilate to bring as much warm blood to the surface of the skin as possible. The photographs below show blood vessels before (left) and after dilation.

We lose heat by conduction, by radiation and by convection. The Schlieren photograph on the far right shows heat radiating from the chest and back, and rising by convection near the throat.

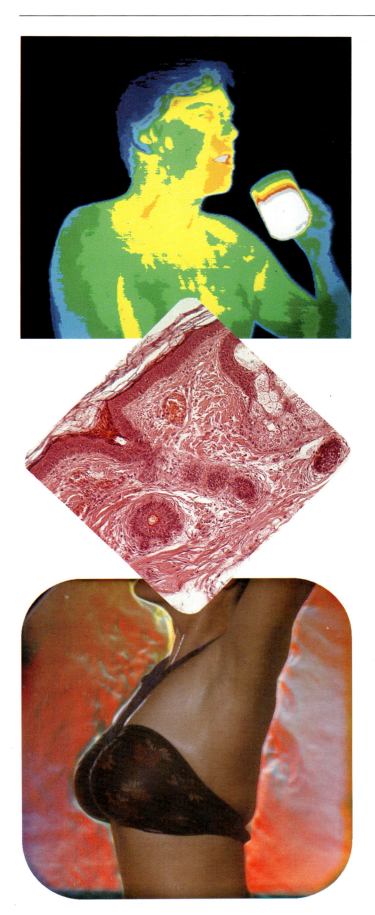

water vapour on your breath, you are still breathing it out, because a certain amount of water secreted at the surface of the lungs is evaporating all the time. Dogs make much use of this by panting in order to lose heat and also lose a lot from the tongue, as they are not nearly so well endowed with sweat glands as humans are.

How successful these mechanisms are depends on the humidity of the surrounding air. If there is a lot of water already saturating the air around us, then the sweat will not vaporize very well. Soon the skin will be soaked with a layer of water that has not evaporated and cooling slows down. This is why we feel warmer in a humid environment than a dry one.

If the surrounding air is dry, on the other hand, there is no physical limit to the amount of evaporation that can occur and we can stand much higher temperatures. In the dry heat of a sauna we can stand temperatures hotter than boiling water – up to 130°C (266°F).

Brain temperature

The result of all these bodily mechanisms is to keep the core temperature within fine limits. But there is also an overriding priority to keep the temperature of the brain at the best level for brain functioning. Temperature changes applied to the head have a much greater effect on heat regulation than the same changes applied elsewhere on the body because of sensitive temperature detectors in the arteries of the neck. The temperature of blood going to the brain can be controlled by the same sort of heat-exchange system we saw in action between the core and surface. Cooling blood flowing from the head down the jugular vein passes close to the carotid artery that brings warm blood to the brain, and the exchange of heat between them can help to prevent the brain overheating. In this way the temperature of the brain can be kept constant, fluctuating less than one tenth of a degree.

This especially sensitive response of the brain to temperature could be the reason that a hot drink such as tea or coffee is actually cooling. It may be that the hot liquid alerts the temperature sensors and sets in motion cooling processes over the whole body, with the overall result that we cool down.

Although the body is remarkably efficient at coping with extremes of temperature, thermoregulation continues all the time, even when the environmental variations are moderate. Through the balancing of heat production and heat loss, our vital functions are protected from the wide variations of the world outside. Because this happens without any need for conscious intervention, our minds and bodies are left free to get on with the business of living.

15 · Action at a Distance

Our lives involve a range of timescales. Days, hours and minutes are all significant in different ways. Some daily events are predictable – the sun sets, the day gets cooler, night falls. Sometimes we have to deal with less predictable events like clouds, wind or rain. On a more personal level too, there are periods of calm predictability . . . and times when life suddenly bursts into action.

The body also has its own 'seasons' – periods of calm alternating with intense biological activity; periods of growth and renewal followed by maturity and decay. All these phases depend for their control on hormones – chemical messengers that can regulate the activities of many different systems in the body.

Like the nervous system, hormones carry signals from one part of the body to another. But nerve signals are electrical, and over in a fraction of a second, while hormones are chemical messengers in the blood which

are active as long as they remain in circulation.

Hormones are at work in our bodies all the time. They help to control many of the body's unconscious activities, particularly those that take place over a long period. There are daily housekeeping tasks, concerned with water balance or digestion, bone growth or heart rate. There are also rarer events like coping with stress or emergencies that suddenly need the collaboration of many different body systems.

Hormones and their effects

Hormones regulate these activities by circulating in the bloodstream to every part of the body. Each hormone is just a molecule of a certain shape. Some, like insulin, are large and complex; others, such as adrenalin, are relatively small. Each shape carries a different message for cells somewhere in the body. It is a message that is as specific as a stop sign, a traffic light, or any other symbolic instruction we find in our world.

The messages deal with how cells work: they tell some cells to start doing something, others to stop, and still others to change their rate of activity. The essential elements are a *gland* which manufactures a *hormone* which then affects a *target cell*.

Almost any type of cell in the body could be a target for one hormone or other. And because the hormones travel in the bloodstream, all cells find themselves visited by hormone molecules even if they are not affected by them. It is only if a particular cell has receptors for a particular hormone that it has any effect.

When an area of the body is bathed by hormones it is rather like the effect of the weather on an area of land.

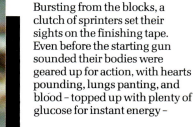

Bursting from the blocks, a clutch of sprinters set their sights on the finishing tape. Even before the starting gun sounded their bodies were geared up for action, with hearts pounding, lungs panting, and blood – topped up with plenty of glucose for instant energy – coursing to the muscles. This enormous change from the resting state is the result of the hormone adrenalin (feathery crystals of this molecular messenger are shown below).

In times of stress our bodies prepare themselves for action – the famous 'fight or flight' response. The brain initiates release of adrenalin from the adrenal glands (situated just above the kidneys) into the bloodstream.

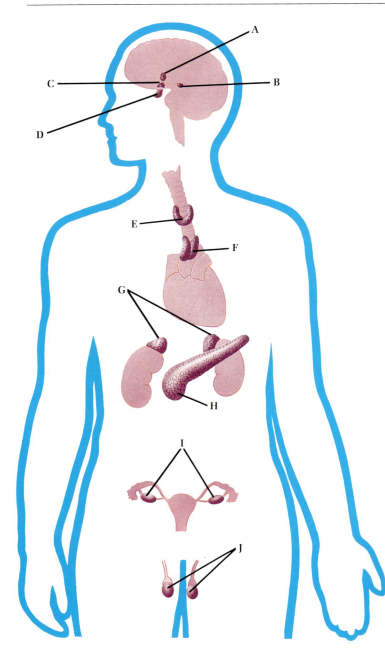

We would not expect a flag to fly each time the sun came out, or a solar panel to respond to the rain. Like the target cells for a particular hormone, flags, solar panels and drainage systems only respond to the relevant type of influence – wind, sun or rain – and, understandably, do not respond to any other type.

The wind does not always blow at the same speed, or the sun stay constantly in the sky. These depend in turn on other factors like temperature, humidity, rotation of the earth and so on. In the same way, the body's 'internal climate' at any one time depends on many different hormones being secreted from many different glands. Fluctuating waves of chemicals pour into the blood from many different parts of the body and are carried in the circulation to bathe every single cell.

From head to groin, there are about ten main glands that produce hormones. The pituitary at the base of the brain secretes a rich mixture of hormones which control some of the others; the thyroid in the neck controls the rate at which we burn our food; the stomach itself secretes hormones to control digestion; the hormones of the pancreas continue that job; the adrenals produce hormones with a wide range of effects on many different systems; and our reproductive life depends on another pair of glands – the testes in men and the ovaries in women.

Like any communication system, the messages carried by the hormones are changing all the time, as the needs of the body change. At different times of day the body has different needs. By deciding when to eat, we impose a timetable of digestion on the body that is controlled by hormones. By choosing what to eat, we give the digestive system a mixture of components which hormones like insulin then have to deal with. Sometimes we need to be alert, at other times to sleep, and each of these states means that hormones have to instruct different body organs and systems to change their rates of working.

The thyroid, for example, controls the rate at which we use the energy obtained from food. When we are asleep, that rate has to be kept ticking over gently, at a level just sufficient to provide energy for those cells and systems that are still at work. If it secreted too little hormone, we would not burn our food at a fast enough rate; we would become fat and sluggish, and would be sensitive to the cold because we could not replace the lost heat fast enough. If the thyroid secreted too much hormone, on the other hand, the body would burn up too much food, producing too much heat; we would then become thin and be very sensitive to high temperatures.

Among the glands that produce hormones or hormone-like substances are the thalamus **A**, pineal body **B**, hypothalamus **C**, pituitary **D**, thyroids and parathyroids **E**, thymus **F**, adrenals **G**, pancreas **H**, ovaries **I** (women) and testes **J** (men).

The hormone-producing cells in each of these glands cluster together around blood vessels so that when their hormones are required they can release them directly into the bloodstream for maximum speed of action.

On the opposite page is a much magnified view of the hormone-making cells inside the thyroid gland. The large spaces between the cells contain a clear, semi-fluid mixture of protein molecules that are the precursors of the thyroid hormones thyroxin and tri-iodothyronin. However, to manufacture them the thyroid glands need iodine. If iodine is lacking in the diet, the glands enlarge in their attempt to remedy the deficiency. The result is a condition known as goitre, shown inset.

Regulating hormone output:
the hypothalamus/pituitary connection

To prevent either of these dramatic alternatives, the output of the thyroid is monitored closely by the pituitary which in turn is monitored by the hypothalamus. If the level of thyroid hormone drops too much, releasing factors are sent from the hypothalamus which stimulate the pituitary to produce thyroid stimulating hormone. The target cells for this hormone are in the thyroid gland itself. As soon as they feel the influence of the pituitary, they step up their production of thyroid hormone; the hypothalamus then notes the increase and slows down its own releasing factor production. Between the three of them, they balance the output of thyroid hormone so that it will give the body just the rate of energy production that it needs for each moment, low during sleep, higher when we are awake and active.

As well as controlling output from the thyroid, the hypothalamus and pituitary gland play a role in controlling many other hormones in a similar way. This type of control system is called 'negative feedback' and is found in several physiological processes.

The hypothalamus and the pituitary are like an international border where nerve signals from the brain arrive to be translated into the language of hormones.

This diagram shows the 'lock and key' theory of hormone action. Three glands produce three different hormones. These travel in the blood to target cells where each slots into its special receptor site. Here the cubes affect the cell membrane, enabling the cones and cylinders to pass through and bind to different molecules inside the cell.

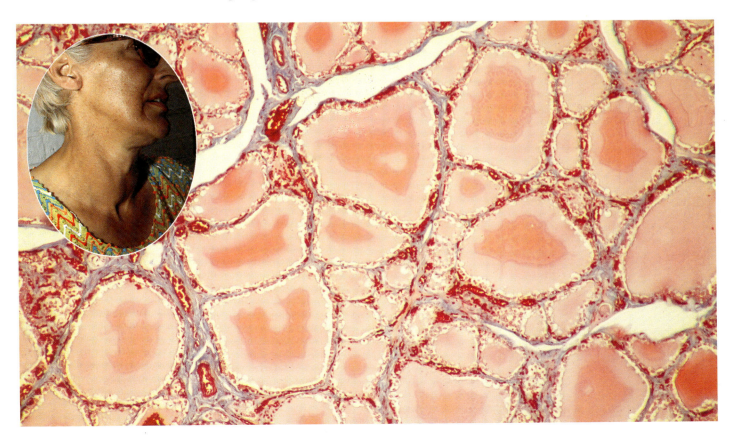

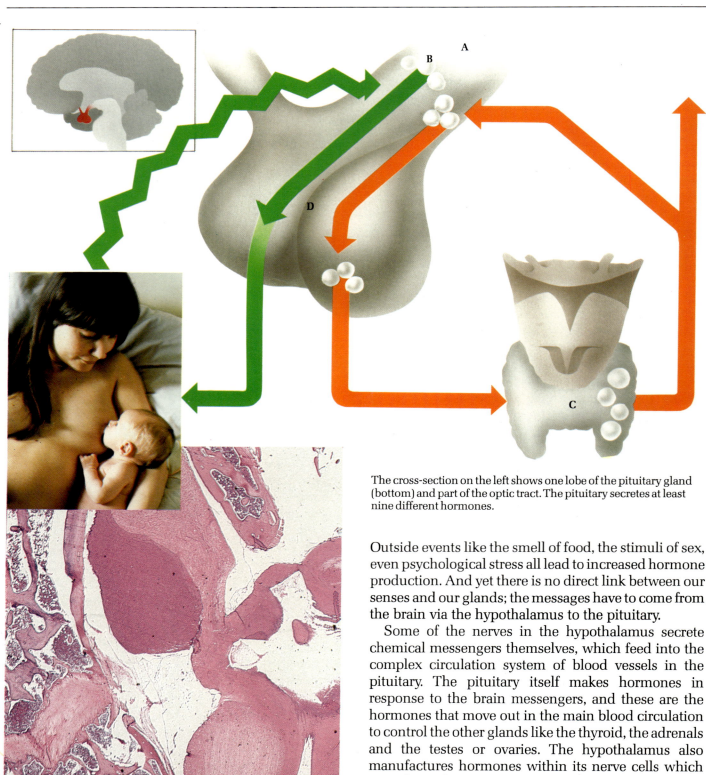

The cross-section on the left shows one lobe of the pituitary gland (bottom) and part of the optic tract. The pituitary secretes at least nine different hormones.

Outside events like the smell of food, the stimuli of sex, even psychological stress all lead to increased hormone production. And yet there is no direct link between our senses and our glands; the messages have to come from the brain via the hypothalamus to the pituitary.

Some of the nerves in the hypothalamus secrete chemical messengers themselves, which feed into the complex circulation system of blood vessels in the pituitary. The pituitary itself makes hormones in response to the brain messengers, and these are the hormones that move out in the main blood circulation to control the other glands like the thyroid, the adrenals and the testes or ovaries. The hypothalamus also manufactures hormones within its nerve cells which extend into the pituitary. In this case, the pituitary acts as a release point rather than a factory. So, like the conductor of an orchestra, the pituitary gland organizes the complex symphony of the body's daily activities.

The events of reproduction are a good illustration of how the pituitary can control events over many different timescales. The ability to bear children takes about 13

In the diagram on the left a sucking baby demonstrates positive feedback (the green arrows): the mother's nipple sends a nerve signal to the hypothalamus **A**, which releases oxytocin hormone **B**, which stimulates milk release. As soon as the baby stops sucking, nervous and hormone signals cease.

The red arrows demonstrate negative feedback between the thyroid glands **C** and the hypothalamus and pituitary **D**; in this case it is over-production of hormones by the thyroids that switches off the production of hormones in the hypothalamus and pituitary which stimulated the thyroids in the first place. Under-production in the thyroids sets in train the brain secretions necessary to step up production again.

The photograph above right shows a section through the thymus gland. The less darkly-staining cells make hormone-like substances that stimulate production of lymphocytes.

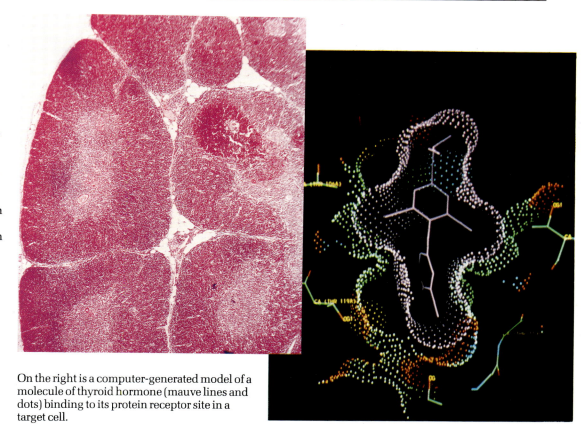

On the right is a computer-generated model of a molecule of thyroid hormone (mauve lines and dots) binding to its protein receptor site in a target cell.

years to develop; menstruation happens every month; pregnancy takes nine months from conception till birth; the womb contracts forcefully over a few hours during a birth; and the events of suckling occur rapidly after birth. None of these is an action we can control consciously with the nervous system – they all depend on hormones.

One example shows how the hormones take effect. When a mother breastfeeds her baby, the first event is the sensation of the baby's lips at the nipple. The resulting sensory message arrives at the hypothalamus and triggers the release of a hormone, oxytocin, from the pituitary. This travels in the blood to the milk glands in the breast and makes them contract, pouring out the milk they contain. As long as there is some oxytocin in the blood, the hypothalamus will sense its presence and cause the pituitary to release more of it. The flow of oxytocin is only disrupted when the baby stops suckling and the hypothalamus no longer receives the touch messages from the nipple. Because the presence of oxytocin leads to an increase, this form of control is known as 'positive feedback'.

With a task like breastfeeding, the job of the hormone is a simple one – it is released in one place, travels to another and has an effect on one specific type of cell – the milk glands. But often the effects of any one

hormone are felt all over the body in many different types of cells. When the whole body needs to respond in a coordinated way with many of its organs and systems, this will often be achieved by means of hormones.

In the days when physical survival was a more chancy business, our bodies developed many emergency systems to make sure that any threat could be responded to as quickly as possible. And hormones played a vital part in those systems. If we follow the moment-by-moment events of a life-threatening situation we can see how one or two hormones command the activities of many different types of body tissue, as part of a general activation of the autonomic nervous system (described in Chapter 13).

Responses to danger – adrenalin and cortisol

The first response to any dangerous situation is, of course, a nervous one. We recognize and interpret the threatening events around us – a fire, perhaps, which needs to be tackled quickly before it spreads. We also know that a fast efficient response is essential. And so our muscles move fast and accurately and the brain plans its next moves.

But there are also long-term problems in a situation of danger. We may have to be prepared for more sustained physical activity and alertness; the heart muscle will

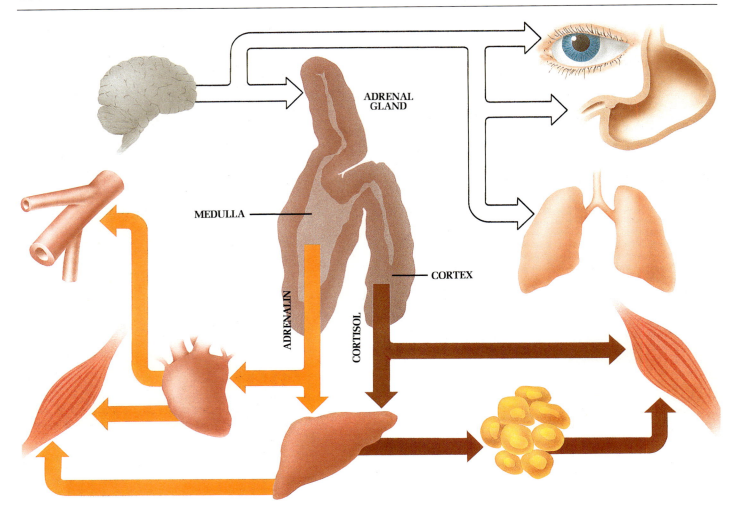

ADRENAL
GLAND

MEDULLA

CORTEX

ADRENALIN

CORTISOL

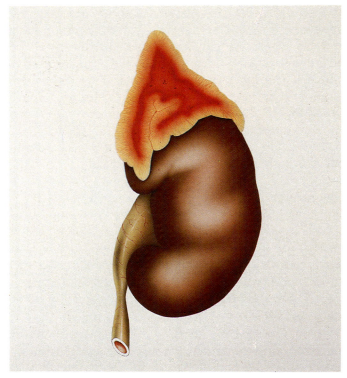

In a contemporary setting the challenge is not 'fight or flight' so much as 'fight or burn'. The dozing firefighters on the opposite page are galvanized into action by the alarm bell.

In the diagram above, the brain perceives the sound and almost at once it has, via the sympathetic nervous system, dilated the pupils for clearer vision, slowed down digestion (giving that familiar stomach-churning feeling) and increased the body's oxygen intake by making the lungs breathe more deeply and forcefully. Almost as quickly, nerve signals flash from the brain to the adrenal medulla and adrenalin pours into the bloodstream. This hormone has several effects (yellow arrows), all resulting in increased supplies of oxygen and energy-giving glucose to the muscles. The heart beats faster and more strongly; glucose stored in the liver (as glycogen) is released for extra energy; blood vessels in the muscles dilate to accept extra blood; and vessels in less essential organs constrict to channel the blood where it is most needed.

By the time the hoses are playing on to the flames the second-line stress hormone cortisol, from the adrenal cortex, is making its effects felt (brown arrows). Cortisol release is triggered by another hormone, sent down from the pituitary. Cortisol encourages the liver to mobilize fat cells and generally to produce more glucose to replace the first flush of adrenalin-released glucose which is being used up. Cortisol also helps break down muscle protein into amino acids, essential for repairing any cells damaged in the emergency.

All this furious activity is mediated by the adrenal glands, which sit bonnet-fashion on top of the kidneys, as shown on the left.

have to be able to contract more forcefully; the body's food stores will be called upon to fuel the extra effort; and the cells in the muscles will have to make the best possible use of the food. If we do not know how long a crisis will last, we have to be prepared for the worst.

All these problems receive help from the adrenal glands, situated on top of the kidneys (which is why they are called 'ad-renal'). Each of them is really two entirely different glands, one wrapped round the other. The core of the gland produces adrenalin, the hormone that characterizes the 'fight or flight' reaction to stress or fear. The outer covering produces another hormone, cortisol, which helps the body to cope with some of the consequences of its fight for survival. Although they are both released into the blood at times of stress, the triggers for the release are entirely different: a nerve signal travels into the gland itself to release adrenalin, whereas cortisol is released when another hormone arrives in the blood from the pituitary gland. But the two triggers have a common starting point – the sudden realization through the senses that life is about to become dangerous.

The effects of adrenalin can be immediate. We often talk of feeling the adrenalin flowing, as the whole body moves into a new state of alertness. The target cells for this hormone are all over the body. In the heart, the muscle cells in the walls contract with greater force and at a higher rate when they are bathed in adrenalin, while the blood vessels supplying the muscles elsewhere in the body dilate to allow more blood to reach them. So as to help the muscles obtain the supply that they need, blood vessels supplying other areas constrict, which is why we sometimes become so pale when frightened – it is a sign that blood is being diverted away from the skin to supply more essential areas. At the same time, direct action of the autonomic nerves makes the passages in the lungs relax so that even more air can reach the lungs to burn the fuel reaching the

muscles. It can also slow down activities that might interfere with the task in hand. Digestion, for instance, is slowed down, and urination is inhibited so that the need to urinate is not a distraction.

To make sure that there is going to be enough fuel for all these hectic activities, adrenalin also mobilizes our carbohydrate food stores in the liver.

Finally, down to the smallest details, it even makes the hairs stand on end, to make us look larger, and it makes the pupils dilate, to allow us to see more clearly.

But while the adrenalin is doing its work, cortisol is also being secreted from the cortex of the adrenal gland. Within minutes of our encountering a stressful situation, our blood contains up to ten times its normal levels of cortisol, which causes a shift to fats as our source of fuel, and helps to build up our rapidly depleting stores of glucose. It also releases extra supplies of amino acids, the repair materials for cells. Finally, it has an anti-inflammatory effect, reducing the sensation of pain; injured firemen and soldiers often do not realize how badly they are wounded until after the excitement of the action is over.

Of course it is not just the *secretion* of hormones that is important – they have to be eliminated as well. It is no good being loaded with hormones from yesterday's crisis. For each of the dozens of hormones, there is a specific chemical process that goes on continuously to break down the molecules in the blood. Depending on the hormone, this process could be over in minutes or continue for hours. Adrenalin, for example, is broken down in the liver in about ten minutes.

At any time, our bodies are full of these chemical messengers, jostling in the blood to carry their individual messages to specific parts of the body. By coordinating many different bodily processes, they help to control the 'seasons' of the body and help us to cope with the storms that arise in the unpredictable world in which we live.

16 · Breath of Life

We live in an ocean of air. We die if we are deprived of it for more than a few minutes. Whatever else we are doing, day or night, the body gives the highest priority to the task of taking in gases from our surroundings. The feeling of urgency that comes upon us if we try to hold our breath for more than a couple of minutes seems a testament to the body's vital need for air. In fact, the most important task is to *get rid of* a gas, carbon dioxide, rather than acquire supplies of oxygen. Both are important, of course, but if breathing stops for some reason, the first danger signals are caused by the build-up of internal pollution rather than the lack of essential supplies.

We have evolved some very elegant mechanisms in the lungs and the tissues to make sure that our bodies are neither starved of oxygen nor suffocated by carbon dioxide. The continuous need for oxygen is not something we are really aware of. And yet, throughout the body, in every cell, there is an incessant need for the gas as part of a process that is essential to life – the production of energy to survive and work. In its end results this process is rather like burning, except that in the body the burning takes place at a much slower rate than in a flame. The heat from a fire can only be released if there is oxygen around to help what is really a fast chemical reaction, and that reaction gives off by-products, especially carbon dioxide, which are not needed and would actually stifle the fire if they were allowed to build up.

Like fire, our body activities use up oxygen, give off carbon dioxide, and create heat. Even in our quietest moments, every cell in the body needs oxygen to stay alive, to renew its own structure, to generate its own electrical signals or to maintain a state of expansion or contraction. To acquire the oxygen it needs it is enough for the gas to be brought very near to it. Then, if there is a higher concentration of oxygen outside the cell than inside, the oxygen will automatically diffuse into the cell. Since we are entirely surrounded by an atmosphere that is one fifth oxygen you might think all our cells would receive their oxygen in this way, by diffusion from outside the body. In fact, when life began on earth all organisms got their oxygen in this way. However, some of our cells are deep inside the body, 10 to 15 centimetres (4-6in) away from the outside, and it would take the gas about a month and a half to diffuse over this

Our remote ancestors took a very important step when they ventured out of the sea on to the land. They acquired the ability to breathe air rather than oxygen dissolved in water. Now, to extend the frontiers of science beyond the ocean of air that envelopes our planet, we take our air supply with us.

distance from the skin to the cells. All the mammals, including ourselves, have evolved a delivery system which takes oxygen in lungfuls and shares it out through smaller and smaller tubes so that it is within easy reach of every cell in the body.

How oxygen reaches the tissues

Every time we take a breath, atmospheric pressure pushes 0.5 litre (1 pint) of air into our lungs. About one fifth of this is oxygen and most of the rest is nitrogen. The way the oxygen eventually reaches the tissues is easiest to understand by following the air through the passages that lead to the lining of the lungs – the thinnest barrier that separates the inside of our body from the outside world. It is taken in through the nose or mouth, drawn down past the Adam's apple and through the vocal cords, and towards a fork where the air passage divides into two, one branch leading to the left lung and the other branch to the right.

After this both branches start to narrow. From a couple of centimetres wide, they divide and then redivide, narrowing all the time until there are many thousands of them, only a millimetre or so in diameter. They end up in the working units of the lung – 300 million spherical chambers called alveoli, clustered like grapes at the far end of each tiny branch.

Encircling each alveolus is a network of blood vessels separated by the thinnest of membranes from the gas in the lung. The vessels are so narrow here that the red blood cells have to squeeze through in single file. This also has the effect of slowing them down from their usual hectic rush around the body and gives them time to come in contact with the oxygen that has arrived in

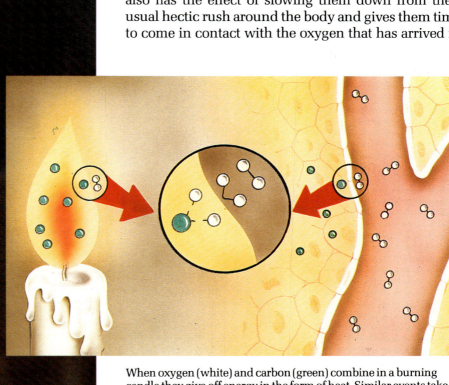

When oxygen (white) and carbon (green) combine in a burning candle they give off energy in the form of heat. Similar events take place in every cell in the body, but much more slowly.

the alveoli. Provided there is a higher concentration of oxygen in the alveoli than in the blood, the oxygen will seep into the red blood cells automatically.

The red blood cell has evolved a special capacity to carry gases around the body and release them when necessary. Each cell is crammed with a pigment called haemoglobin. In the middle of each molecule of haemoglobin are four haem units, containing iron, which have a relationship with oxygen molecules ranging from very friendly to uninterested. When haemoglobin is exposed to an environment rich in oxygen, as happens in the lung, these iron-based components are quite ready to 'shake hands' with any passing oxygen molecule and hold onto it for a while. Indeed, the more concentrated the surrounding oxygen the more avid the haem becomes to take more on board. In the words of Max Perutz, the scientist who has spent his life unravelling haemoglobin, 'Unto haem that hath shall be given; from haem that hath not, even that which it hath shall be taken away.' For as the environment changes on its journey round the body, and there is less and less oxygen outside the cell, the haemoglobin is quite ready to release the handshake and let the oxygen go off to where it is needed.

What makes us breathe?

Since what the red cells pick up in this gas exchange depends entirely on what they find in the lung, the body has to make sure that at all times the lung receives enough air for the job in hand. This is achieved by varying the depth and rate of breathing, the expansion and contraction of the lungs. Only one of these movements requires any effort. We breathe in and let the *lungs* breathe out.

The lungs, like balloons, have a natural tendency to collapse because they are made of an elastic tissue, but they can only collapse so far because they are held in contact with the inside of the wall of the chest by a low-pressure layer of fluid. To take a breath, then, we merely have to increase the volume of the chest for the lungs to be expanded with it. Provided the lungs are open to the air, the pressure will drop and air will rush in. To enlarge the chest, the sheet of muscles called the diaphragm descends at the same time as the ribs are pulled up and out, expanding the chest sideways and from front to back. When the chest muscles and diaphragm relax the elastic tension of the lungs reverses the movement, shrinks the volume of the chest and pushes out the contents.

To keep this movement going for the 600 million or so breaths in an average life, the muscles must receive a signal to contract for each breath. This signal is a nerve impulse from an area low down in the brain. As our lungs reach the right size, another impulse suppresses the first one, the muscles stop contracting and the chest settles back to normal. This happens at regular intervals at a rate – about 16 times a minute – that supplies our oxygen needs while we are resting.

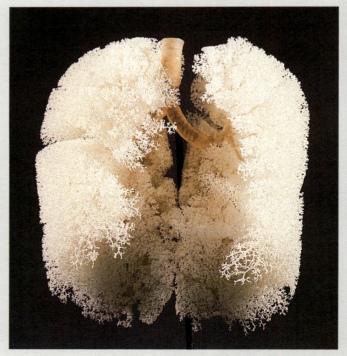

A frog gets most of its oxygen through its wet skin and the blood-rich lining of its mouth – its lungs are very small. For large dry-skinned creatures like ourselves the skin and mouth method of breathing simply will not work, hence the size and complexity of our lungs. The picture above, a resin cast, shows the thousands of air passages in the human lungs.

On the right is a surgeon's eye view of the parting of the ways, the point where the windpipe, or trachea, divides to go to the left and right lung. Note the rings of cartilage visible in the passage on the right. It is absolutely vital that these main airways are held open at all times, unlike the oesophagus which is pressed flat except when food and drink pass down it.

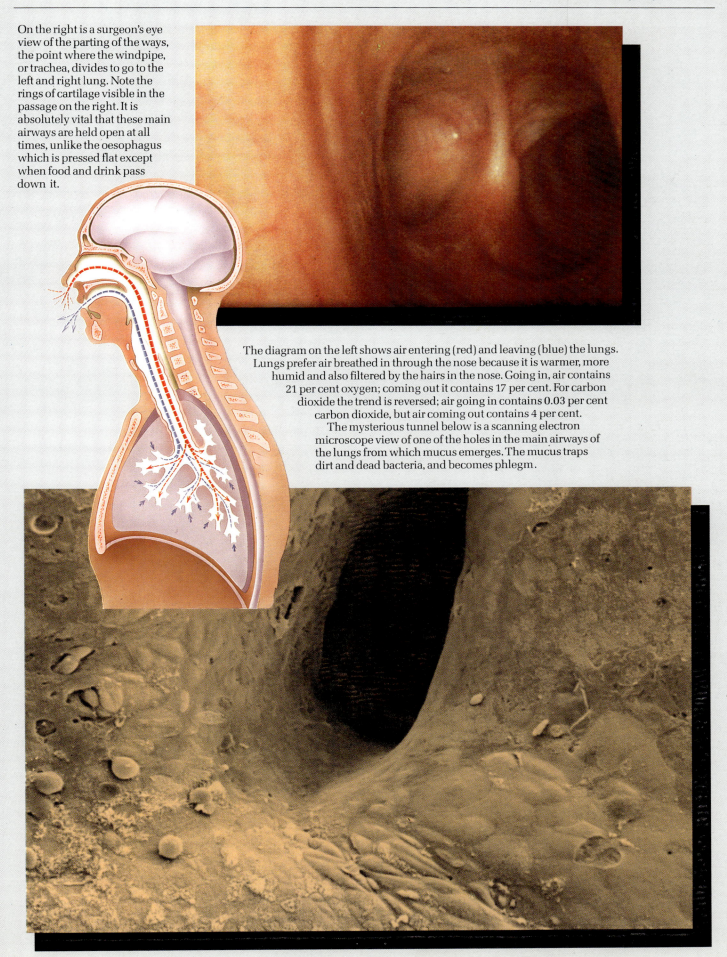

The diagram on the left shows air entering (red) and leaving (blue) the lungs. Lungs prefer air breathed in through the nose because it is warmer, more humid and also filtered by the hairs in the nose. Going in, air contains 21 per cent oxygen; coming out it contains 17 per cent. For carbon dioxide the trend is reversed; air going in contains 0.03 per cent carbon dioxide, but air coming out contains 4 per cent.

The mysterious tunnel below is a scanning electron microscope view of one of the holes in the main airways of the lungs from which mucus emerges. The mucus traps dirt and dead bacteria, and becomes phlegm.

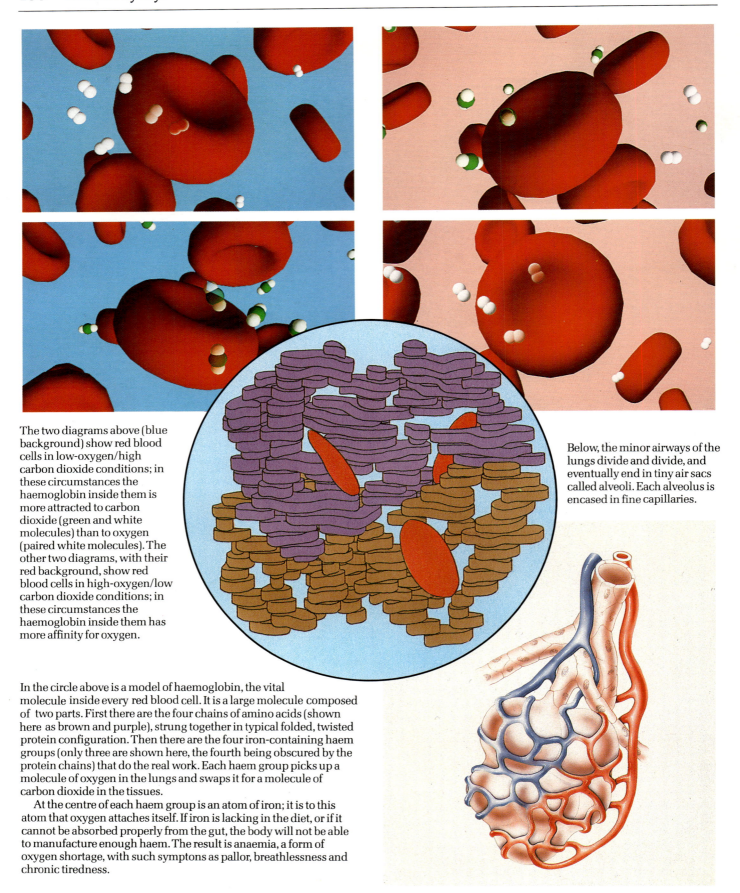

The two diagrams above (blue background) show red blood cells in low-oxygen/high carbon dioxide conditions; in these circumstances the haemoglobin inside them is more attracted to carbon dioxide (green and white molecules) than to oxygen (paired white molecules). The other two diagrams, with their red background, show red blood cells in high-oxygen/low carbon dioxide conditions; in these circumstances the haemoglobin inside them has more affinity for oxygen.

Below, the minor airways of the lungs divide and divide, and eventually end in tiny air sacs called alveoli. Each alveolus is encased in fine capillaries.

In the circle above is a model of haemoglobin, the vital molecule inside every red blood cell. It is a large molecule composed of two parts. First there are the four chains of amino acids (shown here as brown and purple), strung together in typical folded, twisted protein configuration. Then there are the four iron-containing haem groups (only three are shown here, the fourth being obscured by the protein chains) that do the real work. Each haem group picks up a molecule of oxygen in the lungs and swaps it for a molecule of carbon dioxide in the tissues.

At the centre of each haem group is an atom of iron; it is to this atom that oxygen attaches itself. If iron is lacking in the diet, or if it cannot be absorbed properly from the gut, the body will not be able to manufacture enough haem. The result is anaemia, a form of oxygen shortage, with such symptons as pallor, breathlessness and chronic tiredness.

Removing carbon dioxide

Gathering oxygen is only half the job of breathing. The other half of the task is to get rid of another gas, carbon dioxide, one of the waste products of the combustion that is going on in the body's cells. The path travelled by the carbon dioxide is in the reverse direction to the incoming oxygen. Fortunately, the way the blood carries oxygen is mirrored by an ability to carry carbon dioxide along the opposite concentration gradient. In the tissues, the blood loaded with oxygen encounters high carbon dioxide and low oxygen concentrations. Both gases will transfer from a high concentration to a lower one, with the net result that blood swaps its load of oxygen for carbon dioxide. In the lungs, the opposite happens – blood loaded with carbon dioxide dumps its waste and takes on oxygen.

If the cells in the body start working harder than usual, they not only need increased oxygen supplies, they need a faster way to dispose of their carbon dioxide. If it lingers around the muscle cells, for example, it could stifle their attempts to work harder during exercise. This is why the control of breathing depends largely on a close monitoring of carbon dioxide in the blood by the breathing control centre in the brain.

Once activity becomes more strenuous, there are a number of ways in which our breathing pattern can be altered. First of all, any increased muscular activity 'burns up' oxygen faster and, more important, gives off more waste carbon dioxide. Some of this is taken up by the blood and expelled into the lungs. But since these may still be breathing at the same slow resting rate some carbon dioxide stays in the blood. Since the blood bathes all parts of the body it soon reaches the brain, where special sensors will 'taste' the blood, detect increased levels of carbon dioxide, and signal the impulses controlling the breathing to go on for longer, producing deeper breaths. These signals are reinforced by messages from other sensors in large peripheral blood vessels such as the aorta (from the heart) and the carotid artery (in the neck) that can detect decreasing oxygen levels as well as high carbon dioxide. Depending on how much excess carbon dioxide there is, the respiratory centre can also make the breaths come more frequently, so that faster, deeper breaths double or treble the flow of air through the lungs. Another manoeuvre that can help this process is the voluntary pushing out of air from the lungs, so that even if someone who is exerting himself takes the same time to breathe in, he gets rid of the carbon dioxide more quickly and takes another breath sooner.

When intense muscular activity is under way the level of oxygen in the blood adds its appeal to the urgent messages which indicate an increase in carbon dioxide. With extremes of activity, the combination of low

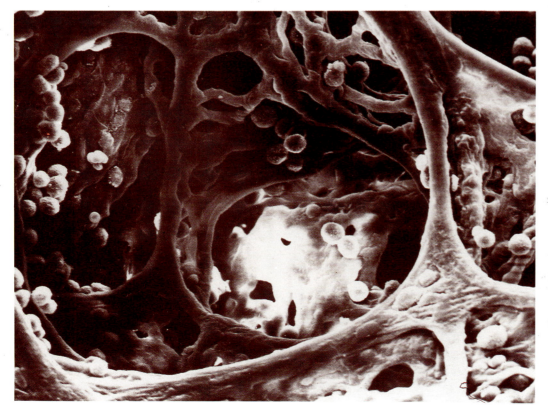

This scanning electron micrograph of the alveoli in the lungs shows that, in fact, they are not separate like grapes in a bunch. Many alveoli have holes in them so that respiratory gases can pass freely between them. Wandering about in these microscopic spaces are special 'dust cells' (alveolar phagocytes) which scavenge and absorb any foreign particles small enough to have reached the alveoli. These cells arrive in the lungs via the capillaries; to get into the alveolar spaces they have to squeeze through the cells that form the capillary walls.

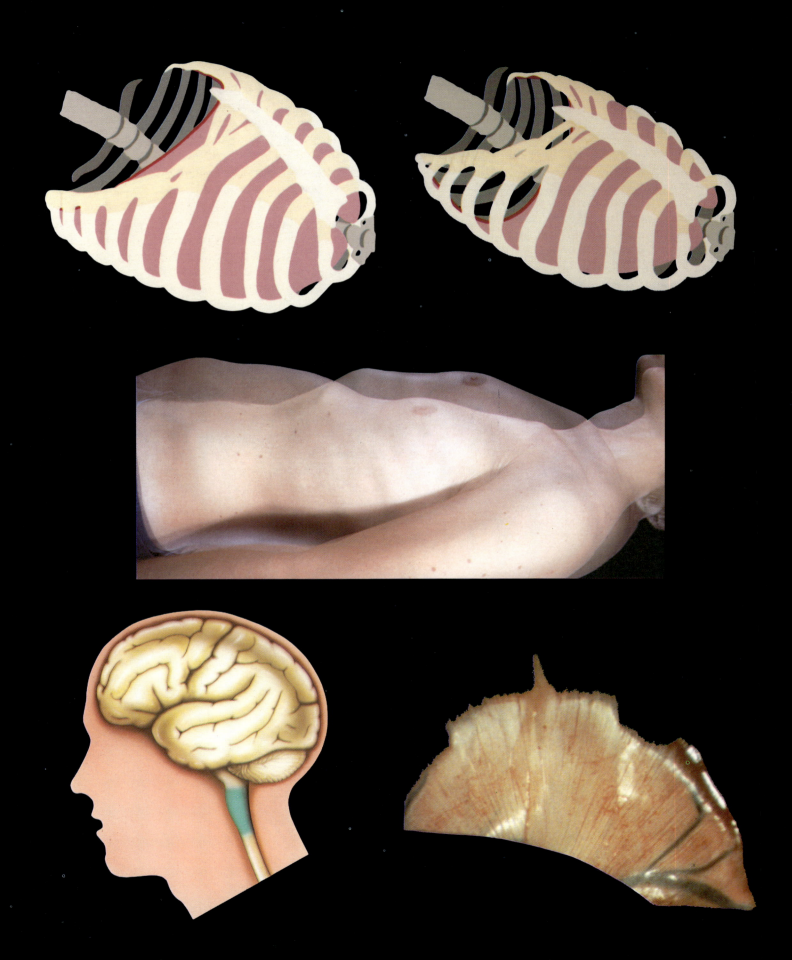

oxygen levels, detected only by peripheral sensors, and high amounts of carbon dioxide lead to the strenuous breathing of the running individual. These impersonal chemical signals in the blood make themselves felt as a gasping craving for air that briefly seems the most important thing in life.

While the *person* concentrates on the task in hand – running fast, for example – the resources of the body are concentrated on supplying the necessary back-up. Even such side-effects of activity as body heat are brought into play. As the muscles work hard they generate heat and raise the temperature. This lowers the level at which the respiratory centre responds to carbon dioxide and means that levels that would not have stimulated any increased breathing at rest will have an effect on the breathing centres of the brain and add to the efficiency of the body machine.

The way we breathe is a very good example of how much the automatic systems of the body can take over and release our conscious brain for the task it does best – decision-making. We *decide* to move faster, and that action brings into play all the necessary back-up systems. Whether we are at rest or physically active, we carry out our daily activities unconscious of the moment-by-moment changes in lungs, blood, brain and cells, which concentrate on the task of delivering the right amount of oxygen for the job we are doing and getting rid of the products of combustion before they can do harm.

The elements of a single breath are shown on the left. The lungs themselves play no active part – all the work is done by the muscles of the diaphragm and ribs.

Nerve messages from the brain stem (shaded green, bottom left) trigger the intercostal muscles to pull the ribs up and out. At the same time the diaphragm (bottom right) flattens, pulling downwards. As chest volume increases, air rushes into the lungs. At this point the chest is inflated and the abdomen stretched. When the muscles relax, the elasticity of the lung walls pulls the ribs in and the diaphragm up, and air is expelled.

Legendary long-distance runner Lasse Viren, 301 in the picture above, wins again. During a race an athlete's air intake may increase to 15 times his resting intake, resulting in at least 50 times more oxygen being delivered to his muscles.

17 · Two Hearts that Beat as One

The heart has become a symbol for many important aspects of our lives – excitement, courage, love and even life itself. It is one of the few organs we can actually feel working inside ourselves, particularly at the more exciting moments in life.

In this chapter we shall discover that the main job of the heart is a straightforward mechanical one – to pump blood around the body – and it does this job with a small number of almost indestructible moving parts. In pushing the blood round the body, the heart makes possible a number of different bodily activities: delivery of food and oxygen to the working cells; distribution of heat, like the water in a central heating system; transport of building blocks for the continual renovation and repair of the body's cells; and the transport of hormones and other essential chemical messengers.

Formation sky-diving is exciting, but not heart-stopping. Even in such an 'unnatural' situation as this the heart keeps pumping and the lungs keep breathing.

The heart's output, which can vary from 6 to 35 litres (10-60 pints) of blood a minute, depends mainly on the rate at which it beats and the volume of blood it expels with each beat. Compared with many man-made pumps, the heart is a marvel of efficiency. Crammed into a space little larger than a human fist, it contains its own power supply, control unit and pistons and can even carry out some of its own repairs.

In our everyday lives, our heart rates go up and down many times a day. Sometimes thoughts and emotions will affect it; at other times it will be the changing patterns of our physical activities. The common factor in all these influences on the heart is the need to supply energy to the body. Sometimes the need is actual and immediate; as when if we run or lift something or use our muscles in some way. Sometimes the heart acts to *prepare* us for possible physical activity, even if we may never need to carry it out. When we are frightened or angry, for example, we prepare ourselves to run away or

fight; and when we are in love, we may well need to prepare for other physical activities.

How blood travels through the heart

Two of the main functions of blood are to carry two gases around the body – oxygen and carbon dioxide. The oxygen is taken in through the lungs and has to be carried to every cell in the body. Every cell gives off carbon dioxide, which has to be carried back to the lungs to be expelled. The blood carrying both these gases travels along tubes, the arteries and veins. To push fluid through tubes some kind of pump is needed.

The human heart actually consists of two pumps. One generates power to push oxygen-rich blood round every part of the body; the other receives the blood once it has been round the body and pushes it through the lungs to get rid of its carbon dioxide and take on more life-giving oxygen.

Pumps do not have to be very complicated. The pumps in the heart are essentially compressible tubes with pulsing walls. The tubes are full of blood, which moves each time the tubes squeeze in. To ensure that the blood does not merely move away from the centre and back again with each contraction there are valves. These act as gates that allow fluid to move one way but not the other. So the pulsing tubes become pumps – liquid coming in at one end is moved on and rapidly replaced by more liquid.

In fact, in a growing embryo the heart starts as two simple tubes. Within three weeks of conception, these are visible in the chest. In another week, the two tubes join and twist until they look much more like the conventional view of a heart, even though they retain their tubular appearance. During the next three weeks, the heart becomes divided up into four chambers. Within six or seven weeks of conception, the cells that make up the embryonic heart start to beat, helping to distribute food supplies from the mother around the tiny body. In a foetus the heart beats quite fast – about twice a second, and the rate does not vary much, as it does after birth, because the foetus has no need to increase its activity.

After birth, however, the heart rate fluctuates to accommodate our bodily needs. From moment to moment, different parts of the body are calling for more supplies of food or oxygen; parts of the body that are heating up because of physical activity need to be cooled, while other parts that are colder than they should be need to be warmed; food in the intestine after a big meal is broken up and needs blood to carry it away

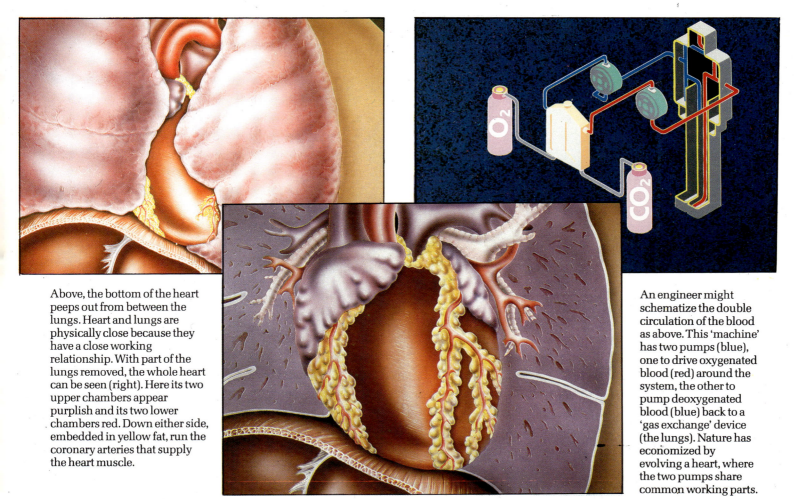

Above, the bottom of the heart peeps out from between the lungs. Heart and lungs are physically close because they have a close working relationship. With part of the lungs removed, the whole heart can be seen (right). Here its two upper chambers appear purplish and its two lower chambers red. Down either side, embedded in yellow fat, run the coronary arteries that supply the heart muscle.

An engineer might schematize the double circulation of the blood as above. This 'machine' has two pumps (blue), one to drive oxygenated blood (red) around the system, the other to pump deoxygenated blood (blue) back to a 'gas exchange' device (the lungs). Nature has economized by evolving a heart, where the two pumps share common working parts.

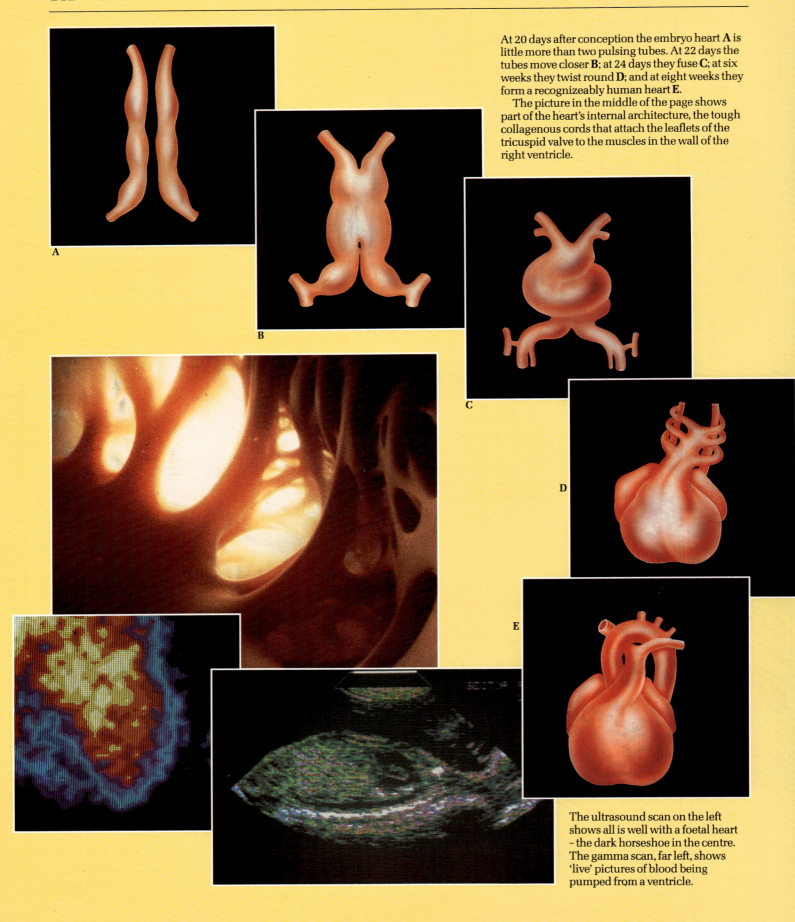

At 20 days after conception the embryo heart **A** is little more than two pulsing tubes. At 22 days the tubes move closer **B**; at 24 days they fuse **C**; at six weeks they twist round **D**; and at eight weeks they form a recognizeably human heart **E**.

The picture in the middle of the page shows part of the heart's internal architecture, the tough collagenous cords that attach the leaflets of the tricuspid valve to the muscles in the wall of the right ventricle.

The ultrasound scan on the left shows all is well with a foetal heart – the dark horseshoe in the centre. The gamma scan, far left, shows 'live' pictures of blood being pumped from a ventricle.

to the rest of the body. In all of these situations the heart plays a significant role pumping supplies to where they are needed, although many other parts of the body help as well.

Although the heart is just two simple pumps that beat as one, the two pumps share a control system and even some moving parts, and so the events of a heartbeat are difficult to separate out as they happen. A normal heart beats at about 70 beats a minute, and when a surgeon looks at it beating in an opened chest it is not easy to see where the blood is coming from or going to. But if we follow one group of blood cells through both sides of the heart, the functions of the separate chambers will become clearer.

The left side of the heart, the first and strongest pump, receives oxygen-rich blood from the lungs. The job of this pump is to give the blood a push that is strong enough to carry it to the furthest parts of the body. The top chamber receives the blood and passes it on to the main chamber called the left ventricle, through a valve. This closes when the ventricle is full, and the muscle in the heart wall contracts, sending the blood out through another valve which leads to the rest of the body.

When it leaves the heart the blood is travelling at a speed of about one metre per second, and it will take less than a minute to travel round the body. Sometimes, in some arteries, it will travel at more than one metre per second, and at other times it will slow down almost to a halt. But within 40 seconds of leaving the heart, the blood will have returned, this time into the right side of the heart, the weaker pump. By now it has given up much of its oxygen and is carrying a large amount of carbon dioxide. The job of the second pump is therefore to push the blood through the lungs, where it can dispose of its carbon dioxide and pick up more oxygen.

Like the left side of the heart, the right has a thinner-walled top chamber that receives blood. In one beat of the heart, the blood rushes from the first chamber through the valve into the right ventricle and on through the other valve towards the lungs. The right ventricle does not pump as forcefully as the left, because resistance in the lungs is lower and therefore the pressure required to pump blood into them is lower.

Once the blood is in the lungs, the waste gas can be exchanged, then breathed out and more oxygen taken on board. The blood then continues back to the left side of the heart to start the journey all over again.

All these events happen almost simultaneously. While the left side of the heart is pumping blood to the working tissues, the right side is pumping to the lungs blood that passed through the left side half a minute or so ago. The same contraction of heart muscle is used to push both sets of blood to their different destinations.

The job of propelling the blood through both sides of the heart is carried out with only three basic components, each of which makes a vital contribution to the efficiency and independence of the human heart. Firstly, there is the solid muscle of the heart walls, which contracts around the blood, reducing the space in the ventricles to a half its normal volume. Then there are the valves, durable flaps of tissue that have to open and shut many millions of times without fatigue or wear. And setting the rate for all this activity is an electrical system, which ensures that the many millions of muscle cells all contract in the right sequence. The activity of these living components is accompanied by pulses of electricity and sound that have become modern symbols of the heart.

Muscle and valves

Heart muscle is a unique tissue and unlike any other muscle in the body. Many of its cells can act individually, contracting on their own without needing any trigger from outside. But if two or three are gathered together, they share a common rhythm which is synchronized by the contact between their cell membranes. When many thousands of them are all in contact, their individual tiny forces add up to the powerful thrust that pumps blood round the body. These cells are great consumers of energy – each cell is crammed with tiny power units called mitochondria. There are many more of these than in other types of muscle and they help heart muscle to extract mechanical energy from fuel more efficiently than almost any man-made device.

There is another useful property of heart muscle that makes it very good at its job. When blood pours into the ventricles, it stretches the walls of the heart. The more the walls are stretched, the more strongly they contract. This helps the normally working heart to avoid having more blood pouring in than it can push out. It also means that when we exert ourselves, our muscular activity returns the blood to the heart in greater volume, and so it is *expelled* with greater force, keeping up the increased blood flow we need.

Heart muscle would, however, be unable to push the blood anywhere without the help of the heart valves. Between the input and the output tubes of the main chambers of the heart, there are four different valves, which flap open easily in one direction, but stay firmly closed to fluid flow the opposite way and they have names derived from their shapes – the mitral valve (like a bishop's hat) and the semi-lunar valve (which has half-moon flaps), for example.

This open-heart illustration shows the electrical control and conducting system of the heart. **A** is the left atrium, **B** the left ventricle, **C** the aorta, **D** the pulmonary artery (blue), **E** the right atrium, **F** the right ventricle, **G** the pacemaker, and **H** the atrioventricular node.

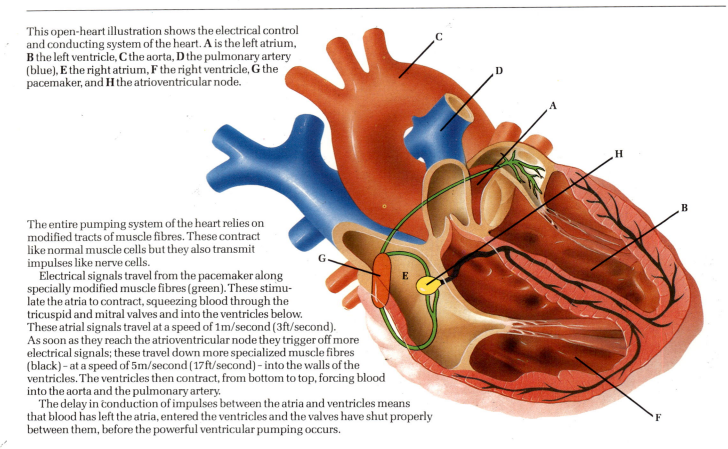

The entire pumping system of the heart relies on modified tracts of muscle fibres. These contract like normal muscle cells but they also transmit impulses like nerve cells.

Electrical signals travel from the pacemaker along specially modified muscle fibres (green). These stimulate the atria to contract, squeezing blood through the tricuspid and mitral valves and into the ventricles below. These atrial signals travel at a speed of 1m/second (3ft/second). As soon as they reach the atrioventricular node they trigger off more electrical signals; these travel down more specialized muscle fibres (black) – at a speed of 5m/second (17ft/second) – into the walls of the ventricles. The ventricles then contract, from bottom to top, forcing blood into the aorta and the pulmonary artery.

The delay in conduction of impulses between the atria and ventricles means that blood has left the atria, entered the ventricles and the valves have shut properly between them, before the powerful ventricular pumping occurs.

Sometimes a valve becomes damaged by disease. When this happens blood can leak back the way it came and the heart becomes less efficient. The mitral valve lets blood in to the left ventricle, but if it fails some of the blood will drain back into the lungs and the working tissues of the body will receive less oxygen. It can be replaced by an artificial valve. Although artificial valves are very efficient, they are never as robust or durable as natural ones.

It is the heart valves that produce the familiar double heartbeat sound, known to doctors as 'lub' and 'dup'. The noise is actually caused by the turbulence that their closing creates in the blood. By listening carefully to the sounds, it is possible to detect whether there is a defect in any of the valves preventing them from closing completely.

When we are at rest, our hearts beat regularly, at a rate of 60 to 70 beats a minute. Every heartbeat is a careful collaboration between the walls of the four chambers of the heart and the four valves. If one part of the muscular wall contracts out of step with the rest, due to damage, or if one of the valves fails to operate the whole smooth running of the heart can be disrupted.

Electrical control: the pacemaker

To coordinate the pumping of the heart, there is a system of electrical signals that spreads throughout the heart muscle and, in fact, can be detected all over the body. Electrodes on various parts of the body show heart activity as a regular electrical pattern, which is called the ECG, the electrocardiogram. Each peak or dip on the ECG marks one element in the heart's cycle of activity. By recording these peaks and dips one can detect conditions that interfere with the conduction of nervous impulses through the heart, or one can monitor the variations in electrical activity of a healthy heart under various exercise conditions.

Each contraction is triggered electrically by the pacemaker, a small cluster of cells embedded in the muscle near the top of the heart that produces regular electrical impulses.

From the pacemaker an electrical impulse travels through the muscle fibres of both atria, conducted from fibre to fibre, until it reaches another specialized group of cells, the atrioventricular node, situated at the junction of the atria and ventricles. These cells act as a relay station and send the impulse onwards down two bundles of fibres running along either side of the wall separating the two ventricles. A ripple of contraction then spreads up through the walls of the ventricles, forcing the blood inside them out of the heart and into the lungs and the aorta.

An open and shut case for the aortic valve, below and right, seen from the aorta. Open, the valve allows blood to surge out of the left ventricle into the aorta. It closes due to back-pressure from blood in the aorta as the ventricle relaxes to suck in more blood.

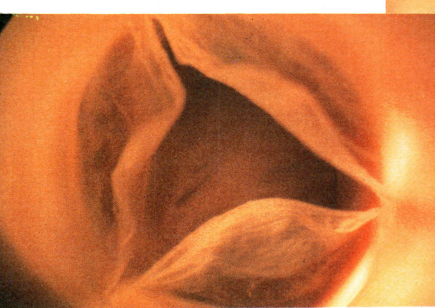

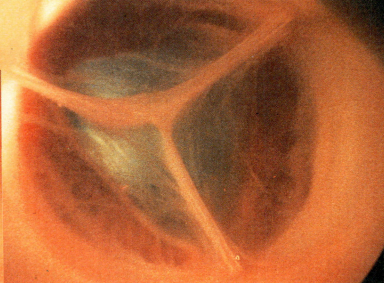

The photomicrograph below shows the oblong, banded fibres typical of heart muscle. Beneath this is a graphic representation of the relationship between the heart's blood flow, the sounds of its valves snapping shut (causing the 'lub-dup' heard through a stethoscope) and the ECG trace of its electrical activity.

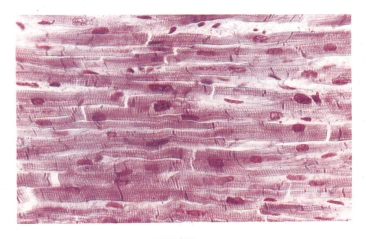

In this way, a regular impulse from the pacemaker leads to an ordered sequence of events in the rest of the heart. Whatever rate the pacemaker sets, the heart follows. But the heart varies its activity depending on the changing situations of life, and can double its rate as its electrical control system responds to signals from all over the body indicating the need for an increased blood supply to various parts of the body.

The rate at which the heart beats is controlled by two independent inputs from the nervous system. These act as an accelerator and a brake, and between them they set the right tempo for the particular physical activity we are performing. Hormones like adrenalin and thyroxin also act as accelerators.

A whole range of signals influence inputs from the nervous system. Some of them are purely the result of physical activity. Stretch detectors in various parts of the circulation monitor the pressure of blood flowing by and, by controlling the nerve impulses, change the heart rate. Other signals come directly from the brain, from the centres that respond to anger, fear or excitement – all emotions which may modify physical activity, and for which we need to be prepared. Hormones such as adrenalin, which are released into the blood in response to the emotional events, have an important effect on the pacemaker.

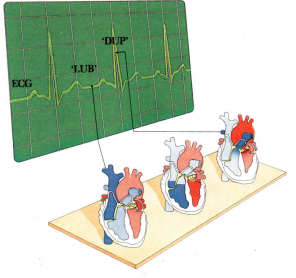

Electrodes placed on the skin (right) can measure heart rate during all sorts of activity. Exercise, fear or excitement can cause the heart to double its pumping rate, sending extra fuel to working muscles or preparing the body for fight or flight.

The coronary arteries

As the heart is virtually all muscle, it is an important consumer of oxygen. At normal working pace, it burns up 10 per cent of the body's oxygen supply just to support its own contractions, even though it accounts for only 0.5 per cent of the body's mass. Although half the interior of the heart is bathed in oxygen-rich blood, little of the oxygen reaches the heart muscle directly. It is only when the blood leaves the left ventricle of the heart that it can deliver oxygen to the heart muscle itself, through its own specialized blood supply system – the coronary arteries.

Just outside the main exit from the left ventricle of the heart into the wide elastic artery called the aorta, there are two unobtrusive openings. When the left ventricle expels its oxygen-rich blood into the aorta on the first stage of its journey to the rest of the body, some of the blood is forced into these two openings, which lead to the coronary arteries, so called because they form a loop like a crown around the widest part of the heart. This is the heart's own circulation, with the job of supplying oxygen to every heart muscle cell.

The coronary arteries are probably the most talked-about blood vessels in the body, because of their crucial role. Just as the whole body stops working if the heart fails, so the heart itself depends on the successful working of the coronary arteries and their associated blood vessels. Diets rich in animal fats, and cigarette smoking, can damage them, although some kind of hardening of the arteries seems to be a general consequence of old age. The heart cells need a continuous supply of oxygen to survive. If they are deprived of oxygen for a very short time – just a minute or two – they die, or their coordination system is affected. And if part of the heart loses its power of contraction, especially the ventricles, the whole body's oxygen supply, the heart's included, can grind to a halt.

Many areas of heart muscle depend on one single branch of the coronary arteries for oxygen. If that branch narrows, because of fat deposits or blood clots, blood can still reach the muscle cells but may not be enough to cope with sudden demands for increased activity. That is when the struggling heart muscle sends the pain signals of angina. If a branch of an artery is suddenly blocked by a blood clot carried in the circulation, no more blood gets through and that part of the heart muscle dies, sometimes throwing the whole heart into chaos from which it may not recover – a fatal heart attack.

If we are lucky, and careful, our hearts may beat 3000 million times before the day when we put too much strain on this double pump and it fails us. Nonetheless, it is more durable than any man-made device. Its efficiency is largely the result of its versatile control system, its ability to beat on its own, and the fact that it has few moving parts.

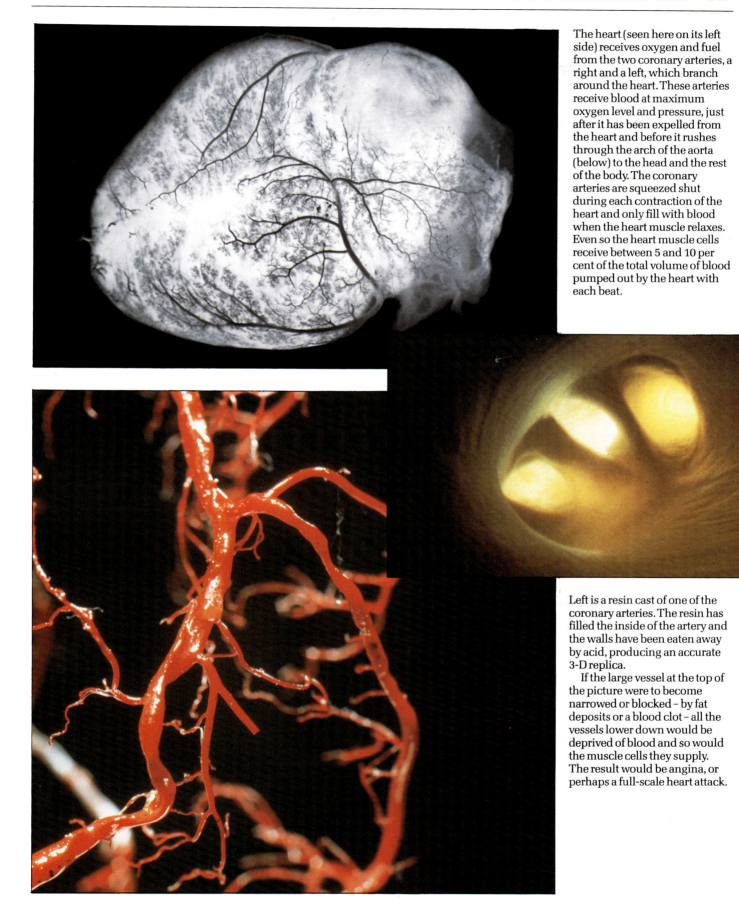

The heart (seen here on its left side) receives oxygen and fuel from the two coronary arteries, a right and a left, which branch around the heart. These arteries receive blood at maximum oxygen level and pressure, just after it has been expelled from the heart and before it rushes through the arch of the aorta (below) to the head and the rest of the body. The coronary arteries are squeezed shut during each contraction of the heart and only fill with blood when the heart muscle relaxes. Even so the heart muscle cells receive between 5 and 10 per cent of the total volume of blood pumped out by the heart with each beat.

Left is a resin cast of one of the coronary arteries. The resin has filled the inside of the artery and the walls have been eaten away by acid, producing an accurate 3-D replica.

If the large vessel at the top of the picture were to become narrowed or blocked – by fat deposits or a blood clot – all the vessels lower down would be deprived of blood and so would the muscle cells they supply. The result would be angina, or perhaps a full-scale heart attack.

18 · The Flow of Life

Every part of the body depends on a continuous flow of blood. The blood is carried in large pipes and small tubes. To move it along these blood vessels, to all the parts of the body that need it, it must be under the right amount of pressure.

To help the distribution system to function successfully, the blood vessels are much more than passive pipes. They have a nerve and muscle structure of their own that helps every part of the body to respond to the changes in demand that happen from moment to moment in our daily lives.

The human body is a community of living cells. Every cell has different needs, for food and drink and energy supplies, and these have to be met by a common supply system. And each cell produces its own waste materials which have to be carried away before they build up to dangerous levels.

These needs are similar to those of a village community, as dependent on water as the body's own cells are on oxygen and food. For the inhabitants of a village there has to be a central source of water pressure – for example the gravitational pull on water in a main tank – that is powerful enough to move the water under pressure to the furthest parts of the village. There have to be channels and pipes to carry the precious liquid, which is used in different amounts depending on the

Every cell of the living body needs blood. Like any other fluid, blood can only circulate in an enclosed system of pipes and tubes if it is under pressure. Just as a water tower provides enough pressure to send water through the pipes of an entire community, so the heart provides the driving force that sends blood to the whole body. Result: water racing through sluices, and blood coursing through arteries.

time of day, the size of the family and the other activities of the village. There are some users who need many thousands of bucketfuls a day, while others have much more modest needs. But even the small users cannot do without. Demand for water and discharge of waste can vary greatly from day to day, depending on the season, the climate and population changes.

In a similar way, the body has to find ways to cope with the varying need for blood during all possible types of human behaviour. Acrobatics require blood in the muscles; brain activity needs blood to be directed to the head; digestion demands blood to be sent to the stomach, and so on. The delivery system has to cope with a continual stream of orders for different amounts of blood to be delivered over different distances, and these orders change from moment to moment. Even a simple action like standing up suddenly changes all the forces that act on our blood. Without some correcting action, the blood would drain to our legs and deprive the top half of our body of essential supplies.

When the blood reaches the parts of the body that need it, it has to be at the right pressure. Too little pressure, and the blood will be useless; too much, and it could damage the cells it is trying to supply.

The most dramatic illustration of what can go wrong in a fluid supply system is a burst water main. If a pipe is fractured or damaged, by age or increased pressure, the results can be chaotic. And in the same way, when a blood vessel bursts in the body, particularly if it is in a vital area like the brain, the blood that rushes through the breach can ruin a life, or end it. Such an event, which we call a stroke if it happens in the brain, can be a result of years of higher-than-normal pressure in the arteries.

On the whole, such massive damage is rare – in a well-cared-for body, the arteries, veins and smaller blood vessels carry out their jobs efficiently for 70 years or more. And this is particularly remarkable because they are not merely passive tubes. We have known for centuries that there are two main types of tubes in the body that carry blood, but the differences were only fully understood once they were seen as the two linked halves of a system design to carry circulating fluid. Exiting from the heart, loaded with oxygen and energy, the blood travels through thick-walled muscular tubes, the arteries, that take an active role in transporting the blood to where it is needed and at the right pressure. Their elastic walls always stretched a little by the blood, the arteries pass on the blood under pressure through

Fainting can be caused by many things, from bad news to standing up too quickly, but the mechanism is the same: a drop in blood pressure temporarily deprives the brain of blood and the body adopts the best corrective posture – horizontal.

Arteries and veins attract the most notice but the real work of the blood is done in the capillaries. The capillaries below are in the gut wall. Oxygen-transporting red blood cells (below) have no nucleus and only stay in active service for about four months.

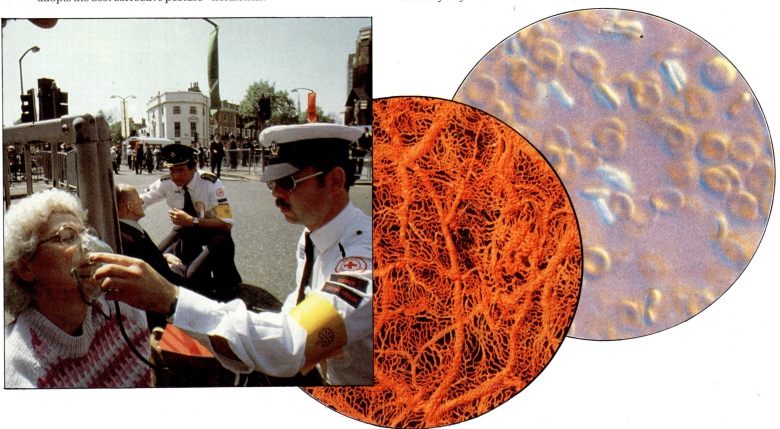

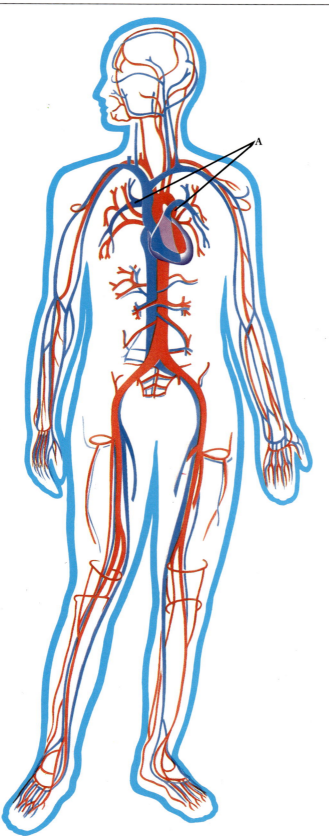

Above is the artery and vein circuit around which, in one month, a single red blood cell may make more than 40 000 journeys.

An artery is defined as a vessel that carries blood away from the heart, and by convention arteries are shown in red because they carry oxygen-rich blood. But the pulmonary artery is the exception; it carries blood away from the heart but the blood it carries is oxygen-poor, on its way to the lungs to pick up fresh supplies. The two branches of the pulmonary artery are labelled **A** above.

successively smaller and smaller blood vessels until it reaches the capillaries, microscopically small vessels, some even smaller than the red blood cells they carry.

Every part of the body has its own capillary system, and it is here that the important work of the circulation takes place. The larger blood vessels are merely a means of transport – it is in the capillaries that oxygen and food pass to the cells and wastes and carbon dioxide pass from the cells into the blood. Once loaded with waste, the blood – now at a low pressure – returns to the heart along distendible blood vessels, the veins.

By following the path of blood through a complete circuit of the body, we can see how well the circulation copes with these changing demands. In its journey to the capillaries, the blood may travel at speeds ranging from 1 metre per second to 2 or 3 metres per hour (3ft per second to 10ft per hour).

Journey from the heart: arteries

One of the fastest stretches of the blood's journey is at the beginning. It is expelled from the left ventricle of the heart up into the body's largest and most elastic artery, the aorta. Like all the arteries, this large pipe has elastic walls, which help to give the blood a good start and to smooth out the flow as it leaves the heart. As the blood

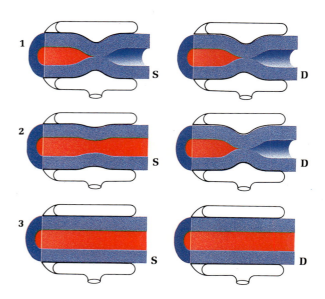

Blood pressure is usually measured in the brachial or arm artery. All that is needed is an inflatable cuff attached by an air tube to a pressure gauge containing mercury, and a stethoscope to hear the surge of blood through the constricted artery during systole.
1 Enough air is pumped into the cuff to close the artery during systole **S**, when the heart beats and pumps blood out under pressure, and during diastole **D**, between beats.
2 Air pressure in the cuff is reduced until systolic pressure in the artery is just enough to overcome it; blood then spurts through, making a tapping sound in the stethoscope and pushing up the mercury in the gauge – usually to a healthy 120mm Hg. With the cuff at this pressure the artery remains closed during diastole, when pressure inside it is weakest.
3 Cuff pressure is reduced still further until it is just lower than the pressure in the artery during diastole, and the mercury level is read off again – it should be around 80mm Hg in a healthy young adult.

rushes out of the heart with the strong contraction of the ventricle, the walls of the aorta expand. Then, when the heart relaxes, the aorta springs back into shape, giving the blood a further push round the bend and off towards the rest of the body. The blood then begins to divide up and flow down a variety of arteries leading to every part of the body. The first organ to benefit from the newly oxygenated blood is the heart itself. As we saw in the last chapter, the heart has its own blood supply to provide the large amounts of oxygen that it consumes. Small passages on each side of the aorta take blood that has just left the heart and distribute it throughout the heart muscle. A little beyond that point, the next important division occurs; two branches of the aorta lead to the brain and two to the arms.

If blood flow in the body drops for any reason, there are systems which try to ensure that all the most important organs are protected. The brain would suffer badly if it was deprived of oxygen for more than a few minutes, and so whenever the survival of the brain is threatened the body gives the highest priority to averting the threat. A little way along the arteries that lead to the brain are pressure detectors, embedded in the arterial walls. Normally these send impulses to the brain when the pressure inside the arteries rises; this inhibits constriction of blood vessels. If the pressure of the blood arriving at these detectors is too low, they send fewer nerve signals to the brain, which then takes

emergency measures to increase blood pressure. It speeds up the heart rate and also shuts down blood vessels elsewhere in the body so that they put up more resistance to blood flow, and so increase the pressure.

The blood pressure sensed by these detectors can be affected by, amongst other things, prolonged standing, which causes blood to pool in the legs. The resulting drop in pressure in the arteries leading to the brain can cause fainting, which can be surprisingly useful; the person usually ends up lying down, which redistributes the blood and makes it easier for the blood to be pumped to the brain. When the pressure rises, the person recovers. Normally, however, a drop in blood pressure is corrected before this drastic stage is reached.

Once the blood has swept past the pressure detectors in the neck, it carries on to deliver its oxygen to every cell in the brain. The blood flow within the brain varies according to the needs of different areas. Parts of the brain concerned with different mental activities need more blood when they are active. Speaking, for example, requires more blood to flow to those parts of the brain that control the voluntary muscles of the mouth and chest, and to the memory areas. Overall, however, the variations in blood flow to the brain in general are remarkably small compared with the much wider variations in the rest of the body.

About one eighth of total blood flow is diverted to the brain while the rest continues to the other areas of the

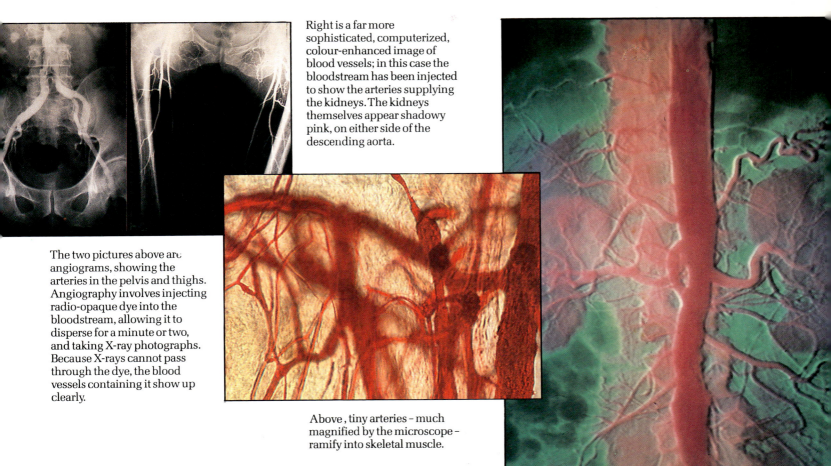

Right is a far more sophisticated, computerized, colour-enhanced image of blood vessels; in this case the bloodstream has been injected to show the arteries supplying the kidneys. The kidneys themselves appear shadowy pink, on either side of the descending aorta.

The two pictures above are angiograms, showing the arteries in the pelvis and thighs. Angiography involves injecting radio-opaque dye into the bloodstream, allowing it to disperse for a minute or two, and taking X-ray photographs. Because X-rays cannot pass through the dye, the blood vessels containing it show up clearly.

Above, tiny arteries – much magnified by the microscope – ramify into skeletal muscle.

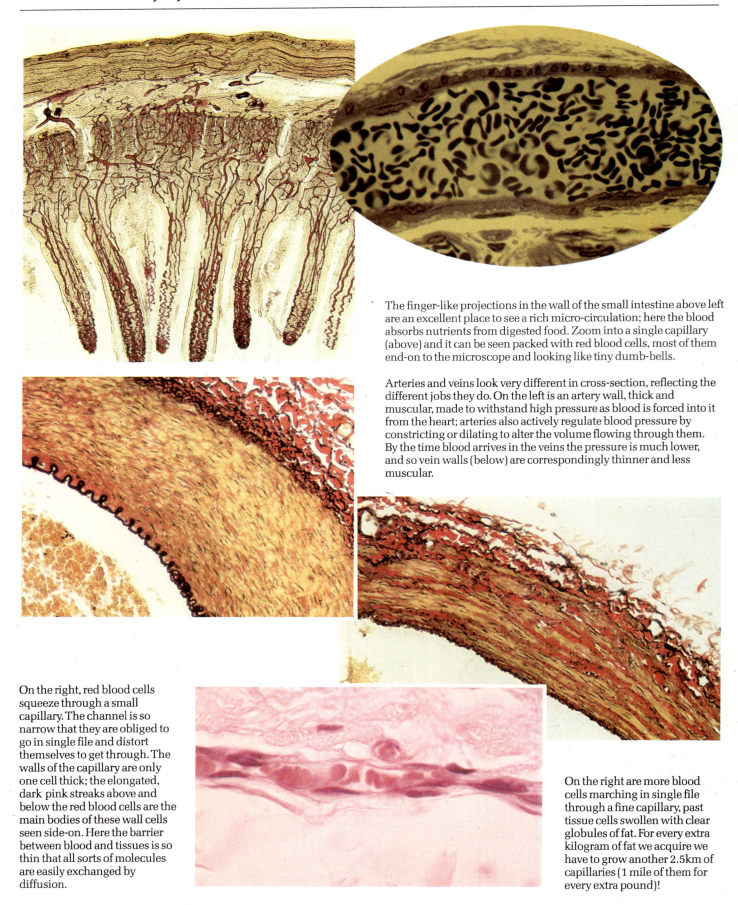

The finger-like projections in the wall of the small intestine above left are an excellent place to see a rich micro-circulation; here the blood absorbs nutrients from digested food. Zoom into a single capillary (above) and it can be seen packed with red blood cells, most of them end-on to the microscope and looking like tiny dumb-bells.

Arteries and veins look very different in cross-section, reflecting the different jobs they do. On the left is an artery wall, thick and muscular, made to withstand high pressure as blood is forced into it from the heart; arteries also actively regulate blood pressure by constricting or dilating to alter the volume flowing through them. By the time blood arrives in the veins the pressure is much lower, and so vein walls (below) are correspondingly thinner and less muscular.

On the right, red blood cells squeeze through a small capillary. The channel is so narrow that they are obliged to go in single file and distort themselves to get through. The walls of the capillary are only one cell thick; the elongated, dark pink streaks above and below the red blood cells are the main bodies of these wall cells seen side-on. Here the barrier between blood and tissues is so thin that all sorts of molecules are easily exchanged by diffusion.

On the right are more blood cells marching in single file through a fine capillary, past tissue cells swollen with clear globules of fat. For every extra kilogram of fat we acquire we have to grow another 2.5km of capillaries (1 mile of them for every extra pound)!

body. Near the beginning of the arteries that lead to the brain are the entrances to the arteries that lead to the arms. It is these arteries that doctors choose when they measure blood pressure. As they are at the level of the heart, the pressure of the blood flowing into either of them is a good indication of blood pressure as a whole. And if *that* is too high or too low it can lead to damaged arteries or insufficient supplies of oxygen and food to the cells of the heart.

The best way to measure blood pressure would be to place a detection device actually inside the arteries. But there is a less drastic way. By squeezing an artery flat with a high pressure until the blood is just able to force its way along it, the maximum pressure transmitted through the blood with each contraction of the heart can be determined. As the blood starts to squeeze through the flattened artery, the turbulent flow of the little spurts of blood is audible through a stethoscope and the pressure of the spurts can be noted.

While some blood gets diverted to the head and the arms, the rest of it flows towards the lower half of the body. The journey is easy at first, down the wide smooth tunnel of the aorta, passing branches off to the spleen, kidney and intestines. The next important junction is a fork from which the arteries lead to the right and left legs. Shortly after this point, the arteries divide again, the main artery of each leg travelling over the brim of the pelvis and a smaller one through a hole in the pelvis. The vessels gradually narrow and the blood slows down as it travels through the thigh and the calf.

In the legs, as in the brain, cells and tissues vary in their need for oxygen and food. Active calf muscles may burn up energy at ten times their resting rate and so the blood flow must increase to supply the needs. The muscular walls of the arteries help to divert blood to the most active sites, just as a community's water supply is diverted by sluice gates and stopcocks to satisfy local increases in demand. The nervous system can control the width of specific arteries, narrowing some and widening others, to regulate the flow.

The capillaries

This same flexibility of supply is found in the smallest blood vessels, the capillaries. Once the blood reaches the feet it is moving relatively slowly, but is still affected by the distant heartbeat. The pressure wave that started with the contraction of the left ventricle of the heart has travelled along the artery and can still be felt as a little pulse at the ankle, just like the pulse in the artery near the wrist. As the blood vessels become narrower and narrower, more and more branches lead off to ensure that every cubic millimetre of tissue has its own blood

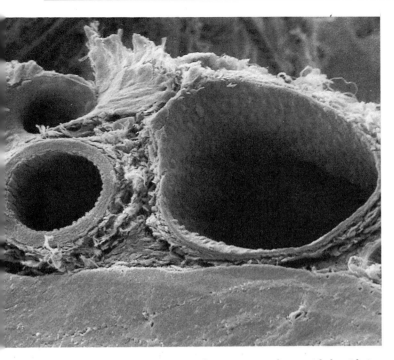

The sectioned ends of a vein and an artery are shown side by side in the scanning electron micrograph above. At any one time more than two-thirds of the body's blood is in the veins, flowing sluggishly back to the heart. That is why the vein above is much wider than its artery companion. The inner surface of the artery is finely corrugated, allowing it to stretch with each surge of blood.

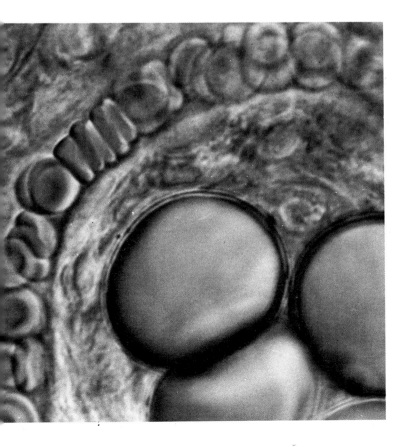

supply. The blood vessels eventually become so narrow that red blood cells have to bend to squeeze through them.

The network of capillaries that threads through all living tissues has its own gateways controlling flow. Around the entrances to these tiny micro-circulations are bands of muscle which can relax or contract to allow blood through or keep it out. The opening and closing of these gateways dictate which cells receive favourable treatment because of their particular needs.

Once they are allowed through, the red cells pass along in single file almost in contact with the tissue cells they have come to supply. Oxygen has been carried 1.5 metres (5ft) from the lungs, attached to these red blood cells. Now they find the opposite conditions to those met in the lungs. With lower concentrations of oxygen in the tissues and higher amounts of carbon dioxide, they discharge their oxygen and take up carbon dioxide – the swap takes place in a second or two.

These tiny vessels, the capillaries, are the furthest point of travel for the blood, the point of supply for all tissues and organs. Every cell in the body is less than 0.2 millimetre away from a capillary and there are over 60 000 kilometres (37 000 miles) of capillaries in one human body. If all the capillaries were open at the same time, they could hold about 6 litres (10½ pints) of blood – more than the total blood volume of the body.

Return to the heart: veins

After it has discharged its oxygen and taken on carbon dioxide, the blood begins its return journey along the veins, back towards the heart and lungs. Although the arteries and veins are similar in shape – they are both just long thin tubes containing layers of muscle and connective tissue – there are certain important differences. Veins have less muscular walls than arteries, but they have valves, which help the blood to pass along them.

Up to this point the blood has travelled easily, helped by the push of the heartbeat. But in the veins, the pressure of the blood has dropped to a fraction of what it was in the arteries, because it has been dissipated while overcoming the resistance of the capillaries.

The body has several methods of helping the blood move upwards, back towards the heart. One of these is in the calves of the legs, where the blood travels very slowly for a moment, until suddenly the vein is squeezed from outside. This squeeze is the result of a contraction of the calf muscles that surround the vein. The body makes use of incidental or purposive leg movements to move blood up to the heart and lungs.

If veins were open vessels like arteries, the blood would move either side of the places squeezed and then move back again. But because veins have valves the

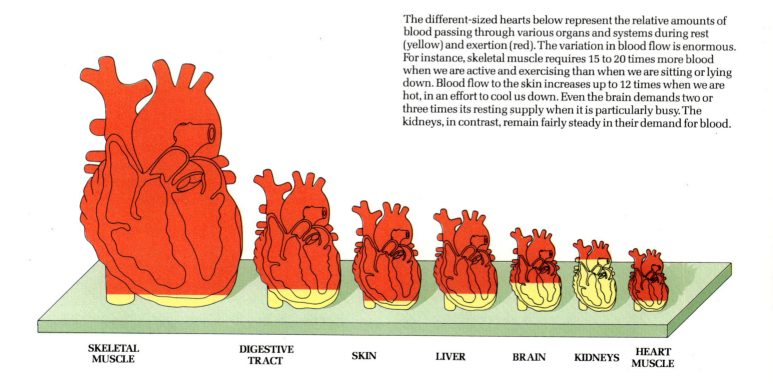

The different-sized hearts below represent the relative amounts of blood passing through various organs and systems during rest (yellow) and exertion (red). The variation in blood flow is enormous. For instance, skeletal muscle requires 15 to 20 times more blood when we are active and exercising than when we are sitting or lying down. Blood flow to the skin increases up to 12 times when we are hot, in an effort to cool us down. Even the brain demands two or three times its resting supply when it is particularly busy. The kidneys, in contrast, remain fairly steady in their demand for blood.

SKELETAL MUSCLE   DIGESTIVE TRACT   SKIN   LIVER   BRAIN   KIDNEYS   HEART MUSCLE

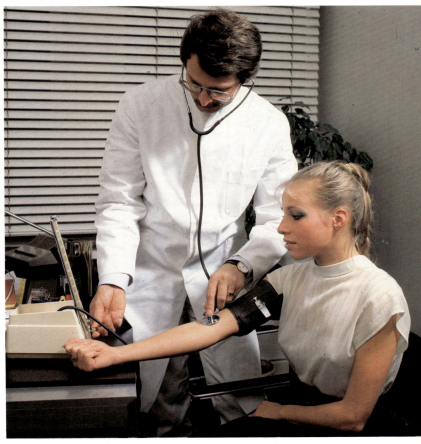

Veins have valves inside them – an open valve is shown on the left. Like sluices in an irrigation system, efficient vein valves are vital to the body's circulation. Monitoring blood pressure is one way of keeping an eye on the body's irrigation system. As the body ages, valves become less efficient, arteries harden and fur up, and the heart has to pump harder.

blood advances. As one valve closes because of the pressure of blood above it, the next valve opens, sending the blood upwards in a surge. Further up the vein there are more valves and more squeezes from the surrounding muscle, which send the blood uphill like a boat going through locks on a canal. This is why people who stay in one position for long periods, standing or sitting, are recommended to contract their calf muscles from time to time to keep the blood flowing properly.

As the blood travels on up the body the veins become wider. Soon the blood is in the large main vein running up the middle of the body, called the inferior vena cava. This is joined by many smaller veins bringing blood from other parts of the body towards the heart. Unfortunately, when this vein reaches the level of the abdomen it is no longer surrounded by the leg muscles that helped it further down, and the blood flow slows

down again. This time help comes from a different source – the muscles used for breathing.

Every time we breathe the diaphragm descends and increases pressure inside the abdomen, squeezing the blood in the vena cava. As the chest expands, the pressure around the outside of the vena cava drops, and blood is 'sucked up' towards the heart. On the way it is joined by the blood returning from all the other organs in the body. With one final surge the blood arrives back at the heart.

This journey happens very quickly: it may take as little as ten seconds during violent activity. And each moment, inside all of us, blood is setting out round the body. The fine control of the ceaseless flow of blood throughout this network of tubes enables just under 6 litres (10 pints) of fluid to deliver a lifetime's supply of oxygen, food, vitamins and minerals.

19 · Accident!

The body is equipped to heal itself without outside help. When bones break or skin is injured, these potentially harmful events trigger the means of their own repair. If the body had not developed this ability, far fewer of us would have survived to adulthood. As we shall see in this chapter, the secret of that repair process lies in the blood. Here we can find the ingredients for the body's own antiseptic, pain relief, bandage and splint.

Although women are relatively familiar with blood as a normal part of life, it is often associated with disaster in the human mind. Most of us first learn that we contain blood as a result of occasional misfortunes like cuts and grazes: injury to any part of the body will allow blood to leak out, showing that it is everywhere flowing just beneath the skin. Although it does not have pleasant associations at such times it does provide the means of our survival after injury, because it travels at speed through every part of the body, transporting the necessary elements of the healing process.

Blood

The blood's role in repairing injured tissue is, of course, only one of its important functions. It is a vital transport system for oxygen and carbon dioxide, and is the main route for hormones to travel round the body and achieve their effects in far-off organs. The closer we look at blood the more we are aware of its complexity. Instead of the thick homogeneous fluid it appears to be, it turns out to be a world of its own, containing a wide variety of types of cells, each with a different job to do.

The various different components of the blood appear to be just a jostling crowd of cells. Some have a nucleus, others not, and some seem to be just fragments of cellular debris. And yet they are all members of the same family; some are more closely related than others, but they all originate from the same type of cell.

Although they all now inhabit the blood, they were not born there. The cells from which they were formed are in the skeleton and the lymph nodes, collections of tissue that we know as the 'glands' that swell up when we are ill. Most of the larger bones have soft specialized areas of marrow beneath the hard exterior, where blood is manufactured and where bone itself is created and destroyed. It is in this dark red honeycomb that the cell division takes place, giving rise to all the different types of blood cell. In fact, it is believed that they can all be traced back to one type of cell, called a 'stem' cell.

Stem cells are rare: perhaps one in 10 000 of the marrow cells is a stem cell, equipped with the potential to produce all the cellular components of the blood. The one cell divides to form two descendents, which stay in the marrow. They each give rise to different families of offspring, some of which also stay in the marrow, while others venture out into the blood stream. One of these cell types breaks up into fragments called platelets, which help later in the clotting process.

Perhaps the most well known and certainly the most numerous of all these descendents is the red blood cell. Although it is descended from cells which have a nucleus containing the genetic material that will be passed on to their successors, the red blood cell is the end of the line. It has lost its nucleus – each cell is just a parcel of haemoglobin, a molecule that is supremely good at carrying oxygen. Because of this, red blood cells cannot reproduce, but are particularly well designed to carry out one job, the transport of oxygen and carbon dioxide in the blood.

In any volume of blood red blood cells hugely out-number other kinds of blood cell. In a sample of 1000 blood cells, most will be red blood cells, one will probably be a white blood cell, and there are likely to be a couple of hundred tiny dots which are the platelets. This really reflects the fact that the main and continuing job of the cells in the blood is transport of oxygen and carbon dioxide. The healing activities we are dealing with in this chapter, which are the responsibility of the white blood cells and the platelets, do not happen all the time. When they are needed, however, there is a remarkable ability for these few non-red blood cells to be in the right place at the right time. Each of them has an important part to play when the surface of the body is injured in any way, providing a whole range of first aid services.

Surface defences and alarm systems

Like any first aid system, the body's response to injury involves such elements as alarm signals, emergency transport, prevention of infection, sealing the wound and providing support. But even before any injury occurs, there is a first line of defence that minimizes any

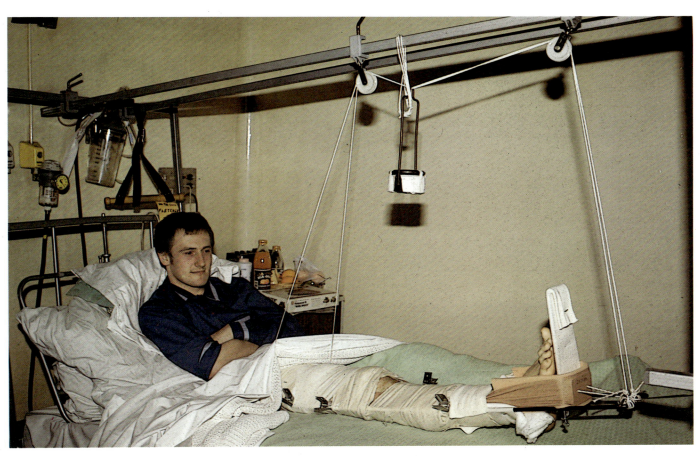

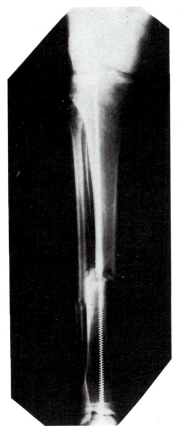
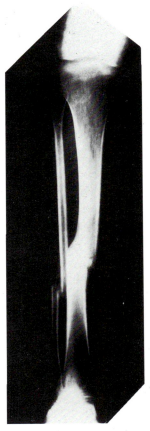
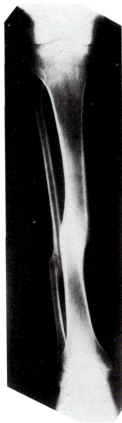

The most common fractures are those that occur in the long bones, in the arms and legs. While these are engineered to withstand stresses acting in the direction of their length, they are vulnerable to twisting and to stresses acting across their length. Treatment consists of positioning the fractured bone in its normal alignment and waiting for the break to knit up. Above, gentle traction is being used to realign broken leg bones.

The fracture of the tibia and fibula on the left has healed without the benefit of alignment; the injured leg is shown immediately after the motorcycle accident that caused it – the zip of the inflatable splint is clearly visible – and again four months and 18 months later. A break as serious as this should have been pinned to ensure that the bones healed straight.

damage that might be done by micro-organisms present on its surface.

The skin is populated by its own colonies of bacteria. We are used to thinking of bacteria as harmful, but that is true of only a small proportion of the micro-organisms that share our world. The skin has a population of beneficial bacteria, which help to defend us in two ways; they keep other more harmful bacteria away merely by occupying areas of the skin that would otherwise be vulnerable, and they secrete compounds that prevent other micro-organisms growing nearby.

Other defences come from the skin itself: several layers of dead cells form a tough barrier against outside invasion, and in the skin there are glands that secrete oils and waxes to trap bacteria. Ear wax, for example, is a sticky grave for many micro-organisms that might otherwise cause ear infections.

In sensitive tissues like the eye there are also glands which secrete a chemical that is lethal to bacteria. In spite of the fact that they provide the ideal conditions for bacterial growth, being warm and moist, the membranes of the eye very rarely become infected because our tears contain an enzyme, lysozyme, that kills bacteria by eating away their cell walls.

Once a breach has occurred, however, help must be summoned. Just as in the outside world we cannot have hospitals and doctors on every street corner and behind every tree, so the body has to be sparing in its internal health resources. Injury is still a rare event, compared with the other activities of the body, and so the cells that are involved with healing patrol the blood stream and visit all the tissues, ready to respond to emergency calls.

When the skin is cut, the dead cells on the surface no longer protect the cells underneath. Bacteria from outside can start to attack and digest the living cells. In doing so, they create chemical by-products. As these diffuse through the tissues they act as a signal that tells the white cells, which may be some distance away, where the trouble is.

There is also a second alarm system, an electrical one, that operates when a living cell is damaged. Normally cells have an electrical charge, and this changes when they are damaged. Some white cells in the blood sense the change and are attracted towards the damaged cells, ready to supply help where it is needed. Sometimes there are so many white cells at the site of infection that they pile up, after they have done their good work of engulfing debris, as pus near the top layer of the skin.

In this way, the repair operation can begin and those inhabitants of the blood that are most effective at dealing with the effects of an accident can be brought to the site of the damage.

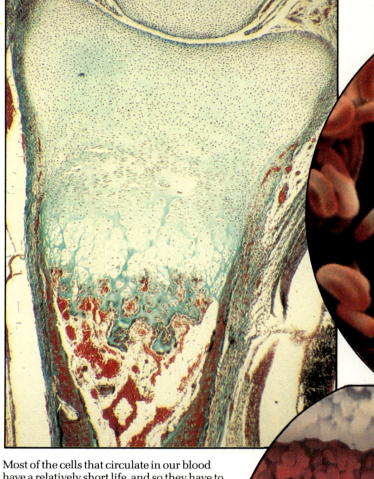

Most of the cells that circulate in our blood have a relatively short life, and so they have to be produced continuously. Red blood cells and various types of white blood cell are produced primarily in the marrow cavities of our limb bones. In the picture above, which shows the lower leg of a foetus, the marrow appears red.

On the right is a close-up of bone marrow, showing thousands of differentiating blood cells at various stages of development. As soon as they are mature they squeeze into the tiny veins that thread plentifully through the marrow and go into general circulation.

Red blood cells, in the circular picture above, are present in huge numbers in the blood – five million in a single pinprick. As their oxygen-carrying days approach, they extrude their nuclei and other organelles. They then become, to all intents and purposes, flexible bags full of haemoglobin, completely dependent on the blood plasma to ferry them round the body.

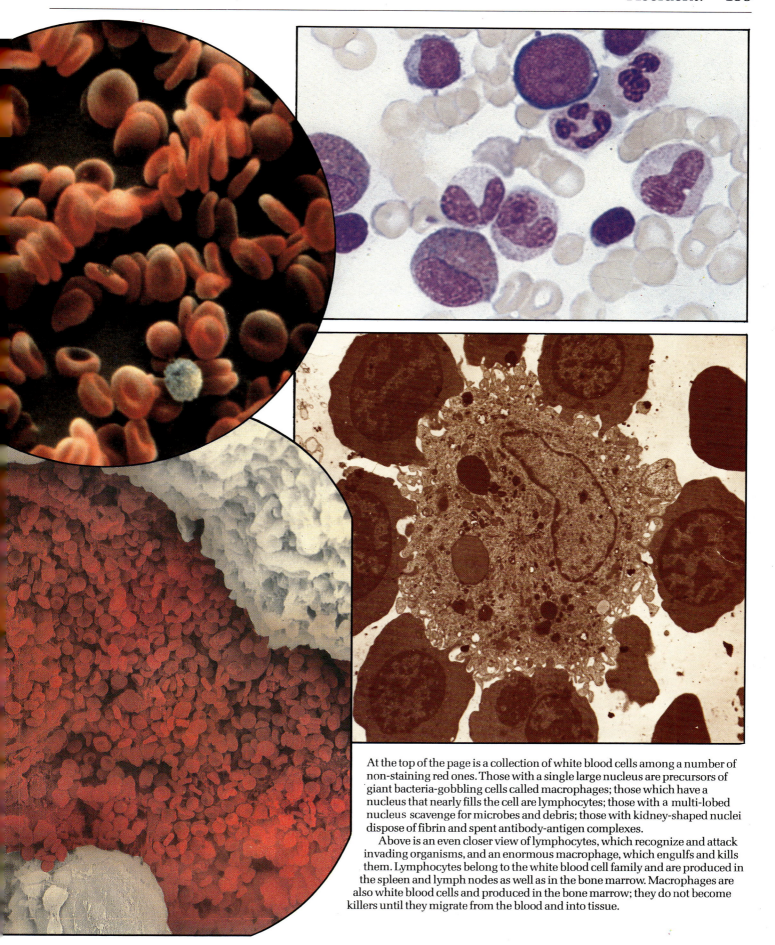

At the top of the page is a collection of white blood cells among a number of non-staining red ones. Those with a single large nucleus are precursors of giant bacteria-gobbling cells called macrophages; those which have a nucleus that nearly fills the cell are lymphocytes; those with a multi-lobed nucleus scavenge for microbes and debris; those with kidney-shaped nuclei dispose of fibrin and spent antibody-antigen complexes.

Above is an even closer view of lymphocytes, which recognize and attack invading organisms, and an enormous macrophage, which engulfs and kills them. Lymphocytes belong to the white blood cell family and are produced in the spleen and lymph nodes as well as in the bone marrow. Macrophages are also white blood cells and produced in the bone marrow; they do not become killers until they migrate from the blood and into tissue.

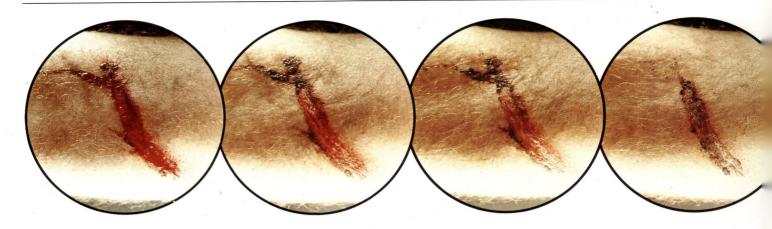

Repairing a cut

If the injury is deep enough to cut into a blood vessel, there must be a quick procedure for closing the gap before too much blood is lost. Here the insignificant-seeming platelets have an important role to play. Any damage to the blood vessels and tissues releases a number of substances, and the platelets circulating in the blood near the cut become sticky as they come into contact with them, and clump together.

The clump of platelets, which also sticks to the walls of the injured blood vessel, forms a temporary plug to stop some of the bleeding. The plug then starts to secrete chemicals which cause the blood vessel to constrict to reduce blood loss, and which also summon up more platelets from further away.

At the same time, a more permanent repair job begins. The damaged tissue has triggered a cascade of activity in the blood that will lead, over the next few hours, to the building of a firm and protective shield over the site of injury. The building materials for this shield are obtained from the blood plasma, the liquid that the cells float in; 'beads' of a substance called fibrinogen are carried in the blood all the time, but are too small to impede blood flow until the chemical actions of clotting lead to the individual beads joining together in long necklaces of a substance called fibrin.

These fibres are used as a basis of a woven mat of tissue that will form a permanent plug for the wound. As they form, they cling to other long fibres and accumulate clumps of platelets. These help to supply bulk for the new plug. They also release chemicals which lead to the contraction and organization of the fibrin so that it eventually becomes a scab that protects the area of the wound and allows new skin to grow underneath. Of course, it is only the first steps of these events that take place in the few minutes after the accident. The full sequence of events will take days – to form the scab and then leave a scar.

The picture sequence above shows the healing of a cut. The first four pictures were taken at 24-hour intervals and the rest over a period of four weeks. Immediately after the injury, the platelets in the blood became sticky and fibrin solidified from the blood plasma; together they staunched the blood and sealed the wound from air and germs.

On the right, much magnified, is a single platelet trapped in strands of fibrin. Blood platelets have no nucleus and only remain in circulation for six to eight days, after which they are destroyed by various refuse-collecting white blood cells. They contain substances which help to constrict blood vessels at injury sites.

While clotting is taking place, white blood cells are heading as fast as they can to the accident site in order to cleanse it. The flesh wound has laid the underlying tissues open to invasion by large numbers of bacteria from outside, many of them finding ideal conditions for feeding and growth on the body's cells. To speed the journey of white blood cells to where they are needed, a sequence of events occurs that we know as inflammation.

First, specialized cells in the tissues, called mast cells, release a chemical called histamine when they are damaged. This has an effect on the walls of the nearby blood vessels, making them leaky, and increasing dilation of them, allowing more blood to reach the damage site. The increased blood flow brings specialized white blood cells called phagocytes, which spread into the tissues in search of clumps of bacteria. When the phagocytes find bacteria, they engulf them and carry them away from the site of the injury, inactivating them on the way and then digesting them with enzymes. They are also able to help clear up the debris remaining after the accident – the dead cells and cell fragments, and dirt.

We are often aware of this increase in activity because of increased soreness around the cut, producing the well-known signs of inflammation – redness, warmth, swelling and pain.

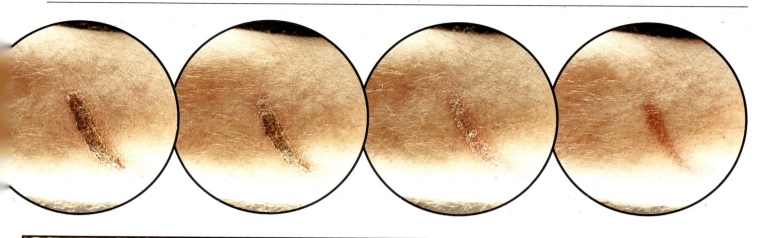

The filaments enmeshing the red blood cells and platelets in the picture below, are filaments of fibrin.

When the skin has fully healed it resumes its flakey, layered appearance – the spectacular landscapes in the two photographs below. The yellow bodies are dirt and bacteria, always lurking on the surface of the skin waiting for admittance. Most of the bacteria that live on our skin are benign; a few, less well adapted, are definitely not.

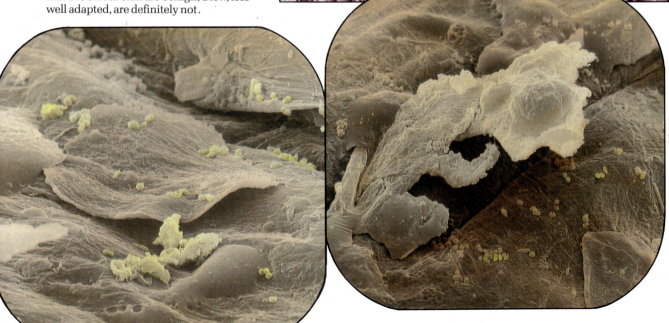

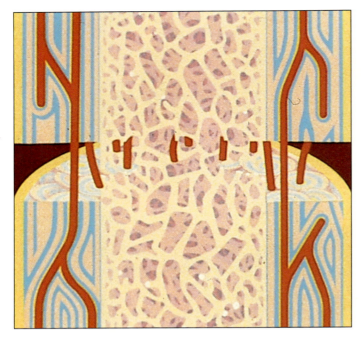

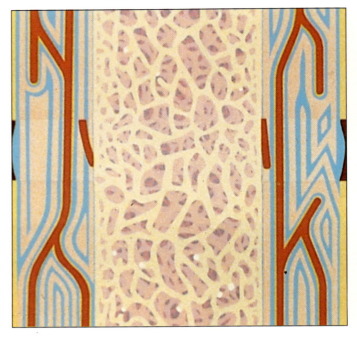

A broken bone has its own repair system, distinct from that which operates in the rest of the body. First, a blood clot forms at the site of the break – blood clots form where there is damage to blood vessels and tissues, whether or not they are exposed to air. Next, blood vessels grow into the clot, as shown in the diagram above. These bring nutrients, raw materials and cleansing white blood cells to the fracture site. Stimulated by the extra oxygen these vessels supply,

collagen-producing cells and cells called osteoblasts (literally 'bone builders') begin manufacturing a matrix that eventually solidifies into bone as calcium phosphate crystals lodge in it. A healing long bone remodels itself according to its pre-injury pattern – with honeycomb bone on the inside and compact bone outside. Usually the only sign of a healed fracture is a slight bulge (blue in the diagram above) round the outside of the break.

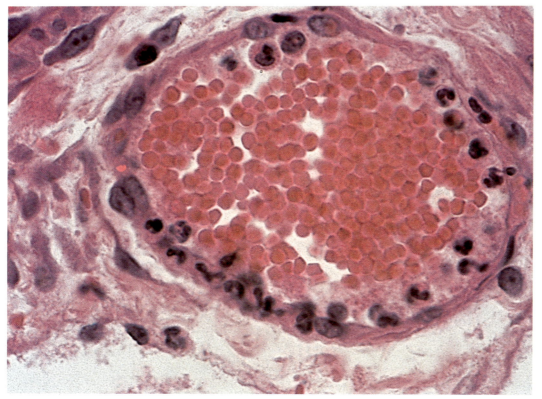

The leaky and distended vein on the left is not malfunctioning; it is helping to repel invading microbes. Under the influence of substances like histamine, produced by mast cells at the site of the injury, small blood vessels become leaky. This allows white blood cells and blood plasma to leak into the surrounding tissues. In this picture, as well as the many red blood cells inside the vein, one can see numerous white blood cells with dark, lobed nuclei gathered just inside the wall. These are preparing to squeeze through it and hunt down bacteria in the tissues outside.

Once all the invading microbes have been destroyed by the engulfing actions of the white blood cells, the final repair work to restore the shape and texture of the injured flesh can begin. Underneath the scab another type of cell, the fibroblast, starts dividing to create more permanent connective tissue, the kind that most of us have somewhere on our bodies, forming the scar that marks the site of a long forgotten cut or injury.

Repairing fractured bones

Up to now we have looked only at tissue damage – the sort of events that happen when the skin is penetrated. But the effects of a violent accident can go beneath the top layer of flesh, to produce broken bones. Fortunately, there is an equally effective bone repair mechanism, also based on the blood, although in this case it is not triggered by the presence of bacteria, since bones can be broken internally, without any contact with the septic conditions of the outside world.

When a bone breaks there is also injury to the tissue surrounding the bone. Blood vessels break and a blood clot forms around the broken ends of the bone. This clot is similar to the one that forms in wounded flesh. It has the same fibrous structure and will provide a suitable nesting place for a new type of cell that will help to remake the hard bone substance. For a day or so, however, the blood vessels contract and actually kill off some of the surrounding bone cells, before the reconstruction work starts, by starving them of oxygen and nutrients.

The clot is next invaded by new cells and blood vessels. One type of cell, the macrophage, removes the debris of the clot. The others are fibroblasts, which in these circumstances can change into two new types of cell to carry out the necessary repairs. One group lays down a framework of supporting skin similar to the skin that forms scars, while the other group produces a callus to bridge the fracture. The gap is then colonized by bone cells which produce new bone. Over several months the callus will be sculpted away by cells called osteoclasts (literally 'bone destroyers') to restore the bone to its original shape.

When an injured human receives the care of doctors and nurses, the most important events actually take place inside his body rather than outside it. Medical care helps to provide the best possible conditions for healing to take place. But in the end, it is not the doctors and nurses who do the healing. It is the body itself.

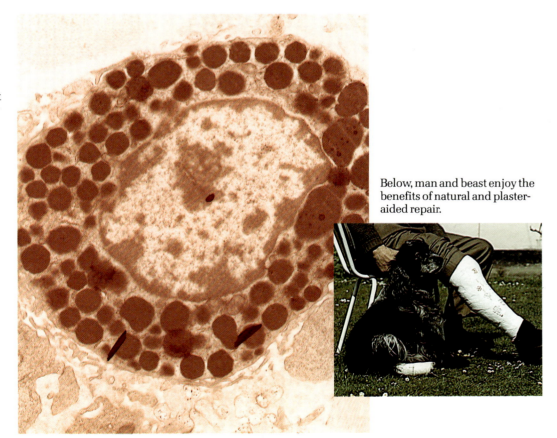

The large cell in the transmission electron micrograph on the right should send shivers down the spine of every hay fever sufferer: it is a mast cell, responsible for making and releasing the potent chemical histamine. The patchy central area is the nucleus; the dark blobs around it contain granules of histamine and other chemicals.

Normally histamine is extremely useful because it causes minor blood vessels to leak and broadcast infection-fighting and toxin-neutralizing cells into injured or infected tissues. We experience this process as inflammation – redness, swelling, itching. But among the stimuli that cause mast cells to release histamine are antibodies (we produce antibodies in response to invaders). Some of us make antibodies against harmless things like house dust and pollen, and our mast cells react by producing histamine – and we get itchy eyes, throat and skin, the symptoms of hay fever.

Below, man and beast enjoy the benefits of natural and plaster-aided repair.

20 · Internal Defences

We have to share our world with a range of other organisms, which are not always as friendly as we would like. Often, the threat posed by these other inhabitants is out of all proportion to their size. An angry bee is troublesome enough, but there is far more danger in the molecular components of the polio virus or the pneumonia bacterium. Bacteria are living cells that are neither plants nor animals; viruses are complex particles which are life-like without being fully alive. Both of them can cause harm to the body in a number of different ways. Bacteria release poisons that damage or kill our cells; viruses gain entry to the interior of our cells and reproduce rapidly, disrupting normal function.

The danger that lies in these forms of attack comes when our first layers of defence are breached. As we have seen, the body has developed an efficient system for reducing the harm that comes from injury when the surfaces of the body are broken and the tissues exposed to bacteria. But sometimes, in spite of this system, foreign micro-organisms succeed in reaching the interior of the body. There they find a warm, moist field, rich in nutrients and biologically useful materials, where bacteria and viruses can create havoc in the orderly running of the body.

Viruses are sub-microscopic organisms which consist of near-naked nucleic acid (DNA). Inert and non-viable on their own, they become dangerously alive and capable of reproducing themselves once they gain entry into living cells. Individual viruses show preferences for certain tissues and locations, but in most cases one infection confers immunity for life. The catch is that some viruses, notably cold viruses, keep one jump ahead of our resistance by literally changing their coats.

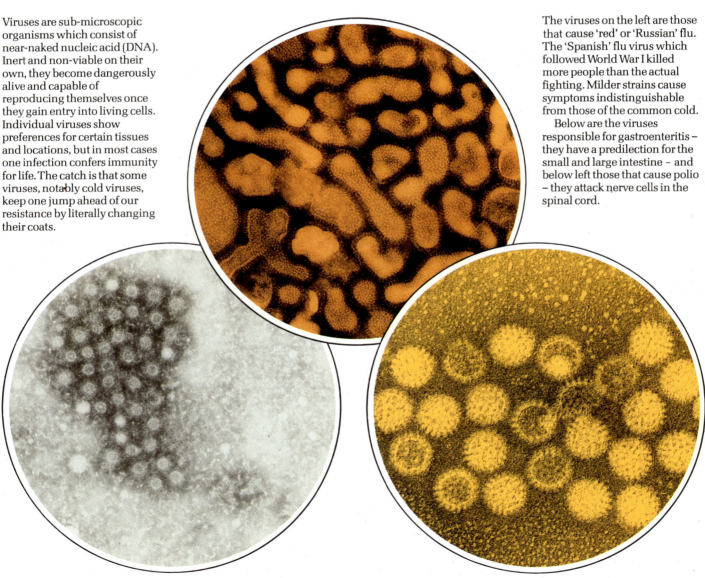

The viruses on the left are those that cause 'red' or 'Russian' flu. The 'Spanish' flu virus which followed World War I killed more people than the actual fighting. Milder strains cause symptoms indistinguishable from those of the common cold.

Below are the viruses responsible for gastroenteritis – they have a predilection for the small and large intestine – and below left those that cause polio – they attack nerve cells in the spinal cord.

Although we do not usually realize it, we are all under attack many times a day. Our bodies are invaded by countless thousands of potentially harmful micro-organisms and yet most of us are healthy most of the time. This is because the body is equipped to deal with every invader in a unique way. It has evolved a system that identifies every possible harmful intruder, and attacks it where it is most vulnerable

Furthermore, the body has a life-long memory for micro-organisms, so that once we have been invaded by a particular type of organism, we can eliminate it even more effectively if it ever attacks us again.

In spite of the useful protective characteristics of the skin – it is tough, waterproof and flexible – there are many weak spots on the body surface where harmful organisms penetrate beneath the surface. Because we need to breathe, reproduce, eat and excrete, we have many vulnerable orifices leading to the body's interior.

Even if there is not direct access to the body's tissues, there are areas with membranes or layers which have to let through some substance from the environment. Air, for example, or food molecules, must pass across the barrier between us and the outside world, and it is at such crossing points that we are most vulnerable to intruders. And the intruders themselves have developed to make the most of these opportunities.

Effects of an attack

If we look at the bodily events that accompany a common cold we can see how they reflect the struggle between body and invader, a struggle which the body usually wins. The common cold virus is well suited to using the weak points in our body to establish a foothold so that it can thrive and reproduce itself at our expense. It is a small particle and there is nothing about its size or shape to suggest the harm it does.

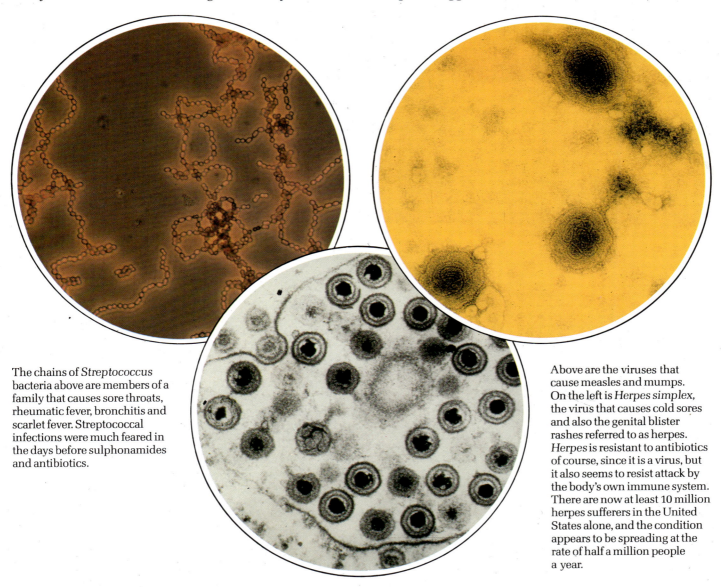

The chains of *Streptococcus* bacteria above are members of a family that causes sore throats, rheumatic fever, bronchitis and scarlet fever. Streptococcal infections were much feared in the days before sulphonamides and antibiotics.

Above are the viruses that cause measles and mumps. On the left is *Herpes simplex,* the virus that causes cold sores and also the genital blister rashes referred to as herpes. *Herpes* is resistant to antibiotics of course, since it is a virus, but it also seems to resist attack by the body's own immune system. There are now at least 10 million herpes sufferers in the United States alone, and the condition appears to be spreading at the rate of half a million people a year.

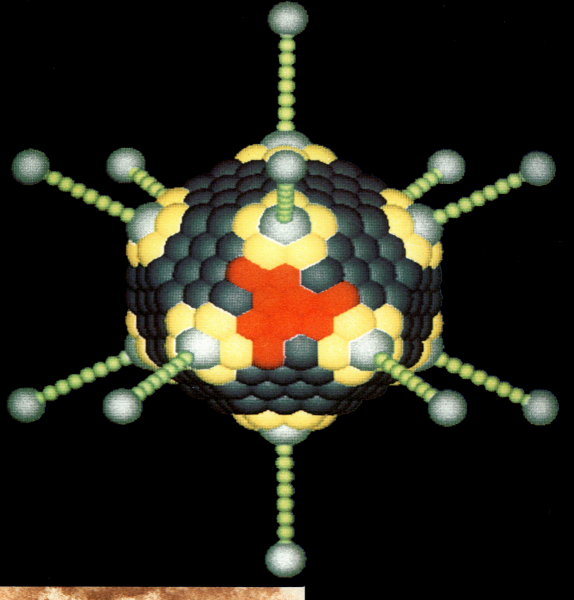

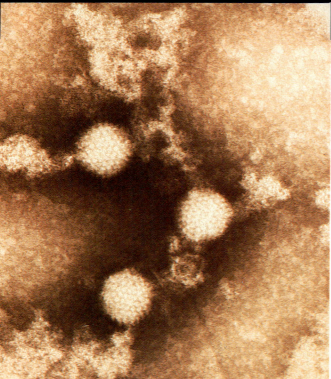

Above and left are two views of the cold virus *Adenovirus:* the image above is a computer model, and the image on the left is a transmission electron micrograph. As well as causing inflammation of the membranes of the nose and throat, *Adenovirus* can cause other respiratory ailments and conjunctivitis.

There are at least 30 different viruses, some more virulent than others, that trigger off cold symptoms. These generally make themselves felt within 48 hours of the virus gaining entry and generally disappear about a week later, which is about the time it takes for our immune system to manufacture the appropriate antibody. Proprietary and folk remedies may alleviate the symptoms but they do nothing to knock out the virus. Antibiotics are no good either, though they may deal with any bacteria that have installed themselves in sites already infected by the virus.

Vaccines, which provide us with ready-made antibodies, are effective against some viruses, especially those that cause smallpox, polio, typhus, rabies, yellow fever, measles and German measles. Thanks to widespread immunization, smallpox is now a disease of the past; the smallpox virus now lives not in the population at large but shut up in a few laboratories.

We usually acquire a cold virus from someone else with whom we have been in close contact. This virus cannot survive for long away from a human being. It will nestle for a while on warm human skin but, unless it can find some cells to invade, within a few hours it will have lost its power to reproduce.

Fortunately for the cold virus, humans are gregarious, and most people who have a cold will end up giving it to someone else. When the virus is transferred from hand to hand for instance, it is likely very soon to be placed conveniently near the tissues it likes best, in the warm moist passages of the nose at about 35°C (95°F). Its aim is to make lots of copies of itself, and it embarks on this task using the component parts of *our* cells as materials. Under the strain of having to house and feed these thousands of virus particles, the normal working of our cells breaks down and they disintegrate, shedding virus into the blood and tissues where it can invade more cells and start reproducing again.

At first, the effects are mainly confined to the area around the nose and are similar to the inflammation reaction described in the last chapter. The blood vessels in the area become leaky and white cells infiltrate the area and try to reduce the damage. Often, this local action is not enough to prevent the virus reproducing itself at a faster pace than can be dealt with locally and the virus spreads in the bloodstream to other tissues.

A similar series of events takes place for hundreds of other harmful intruders, some of which harm us by taking over our cells, while others poison us with chemicals which they secrete once they are in the body. All of them depend for survival on being able to reproduce inside us – all of *us* depend for survival on our ability to identify the intruders and eliminate every one of them as quickly as possible.

One of the signs that the body has been invaded in this way is the swelling up of lymph nodes – collections of tissue that we talk of as 'swollen glands' when we are ill. In the case of an upper respiratory tract infection, for example, a pair of nodes near the throat swell up, showing that somewhere in this region there is an attack being mounted against invaders.

The lymph nodes are linked together as part of a network that runs throughout the body, parallel to the circulation of the blood but separate from it, apart from occasional points of contact in the nodes. This circulation carries a fluid called lymph, which is just like the blood plasma without red blood cells in it. It functions as a drainage system, flowing through every organ in the body and carrying off any bacteria or broken-down cells to the nearest lymph node to be disposed of. It also collects fatty acids from the intestine and empties them into the blood.

The largest individual part of this system is the spleen, which is effectively the lymph node for the blood, designed to carry out the same task of detecting and removing useless or harmful cells. One of its roles is a 'breaker's yard' for old red blood cells. These have a normal life-span of about four months. Towards the end of that time they show some external signs of ageing that are recognized by the spleen, which is therefore able to pick out of the blood flowing through it just those cells that are losing their powers. These are then broken up by the cells in the spleen and the main useful component, iron, is recycled to other parts of the body where it can be used to build new red blood cells and start the process again.

But the main job of the spleen is to help the lymph nodes in their task of dealing with foreign invaders before they can do too much harm. Any organism small enough to penetrate the surface defences will arouse the body's surveillance system in one of two ways. Either it will produce poisonous molecules as part of its normal activity, or its surface will be made up of molecules that are different from any of the body's own components. In either case, the body recognizes the invader as 'not-self' and this recognition alerts the immune system.

In the case of the cold virus, as we have seen, it is the whole virus that does the harm, by swamping the body's cells with its offspring. But other organisms, such as the bacteria which cause diphtheria, wreak their havoc by producing toxins which poison the body's cells.

Production of antibodies

It is fortunate for us that the human body is able, in a few hours or days, to identify the harmful virus or poison and manufacture a chemical, called an antibody, which will attack and neutralize just that component and no other. The antibody is made as a mirror image of the harmful molecules it is designed to attack and therefore can attach itself firmly to a relevant molecule and halt its activities, but has no effect on any other molecule.

The most important elements in this process are white blood cells called lymphocytes – produced in the lymph nodes where they can seep into the blood circulation. These cells respond to a component of the virus or bacteria, such as a molecule in its external coating, and quickly produce vast quantities of exactly the right antibody to overcome the specific invading micro-organisms. Within hours or days of a new attacking organism entering our bodies – one we have never encountered before – our blood contains massive supplies of the antibodies designed to fit onto the harmful part of that organism.

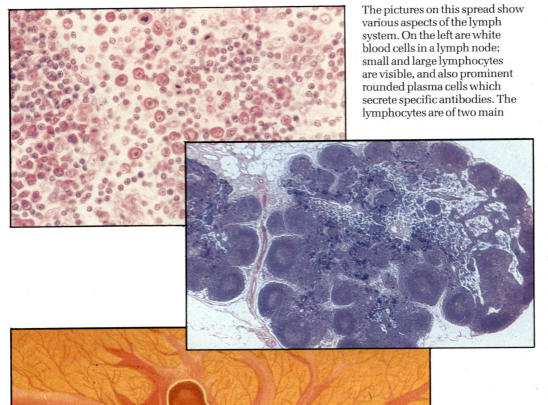

The pictures on this spread show various aspects of the lymph system. On the left are white blood cells in a lymph node; small and large lymphocytes are visible, and also prominent rounded plasma cells which secrete specific antibodies. The lymphocytes are of two main kinds, B lymphocytes and T lymphocytes. The Bs make antibodies and the Ts recognize and attack infected body cells directly. The Ts are also part of the non-stop cellular vigilante system that protects us from cancer – they identify and destroy body cells that have gone 'wild'.

Left is a slice through a lymph node. The blue stain clearly shows the lobed structure of the outer region of the node. In real life a node like this would be about the size of a haricot bean. Blood arrives via arteries and lymph via lymph vessels, and both are filtered and cleaned by the lobe cells.

The red and orange diagram below left shows a section through the spleen, a vital component of the lymph system. The small veins that drain it empty into the splenic vein. In real life the spleen is sandwiched between the stomach and the diaphragm. In addition to producing lymphocytes, it disposes of ageing red blood cells.

The photograph on the right is a close-up of a tiny duct deep inside the spleen. Forming the walls of the duct are networks of cells with many finger-like and veil-like projections. One red blood cell, shown at **A**, is in the act of squeezing into the duct. The cells already in the duct are red blood cells and a lymphocyte, half visible and looking like a fluff-covered ball.

The upper photograph on the far right shows a lymph vessel running parallel to a blood vessel. Valves (like the one just visible as a faint cross in the upper left hand corner of the picture) ensure that the lymph flows in one direction only. Into the lymph, admitted through the very thin walls of the lymph vessels, go tissue debris and particles too large to squeeze into the capillaries.

The other photograph on the far right shows lymph vessels and nodes in the lower abdomen and groin. This is a lymph angiogram, in which the lymph vessels have been injected with a dye that is opaque to X-rays.

But how can the body be equipped to produce very quickly just the right defenders, or antibodies, to deal with thousands of different invaders? Imagine someone needs a dozen or so suits made to fit only him. He can acquire these suits in two ways: he can go to a tailor, who will take all his measurements and cut the materials to produce the suits he needs, or he can go to a large ready-to-wear clothing store and try on the suits one by one until he finds one that will fit him, and then order a dozen copies. The body could work in one of two similar ways. The lymphocytes could 'measure up' the invaders and then design and construct the right antibodies. Or there could be many thousands of different types of lymphocyte, each already equipped with the ability to make just one type of antibody. Then if the invaders encountered the one type of lymphocyte that matched them, antibody production would be triggered and the invaders attacked.

In fact, the body uses the second method. Although it may be difficult to believe, among the lymphocytes that lie in wait in the lymph nodes or patrol every corner of our body looking for foreign invaders, there are thousands of types of cells each able to make only one type of antibody, ready to identify the relevant harmful molecule and destroy or neutralize it.

When a molecule from outside comes into contact with the specific lymphocyte that can match it, the lymphocyte starts to divide and some of its offspring become dedicated to the task of producing as much antibody as possible. These antibody-producing daughter cells are crammed with the synthesizing equipment that will make new antibody molecules at speed and expel them into the tissues.

One surprising feature of this sytem is that every human is equipped at birth or soon after, as it takes time to become fully effective, with the *ability* to make antibodies for all the harmful viruses or bacterial products he is ever likely to encounter. In the blood and tissues of a young child are cells that may not be needed for 20, 30 or 40 years, or may never be needed. They carry cellular machinery that could very quickly produce specific antibody molecules if required.

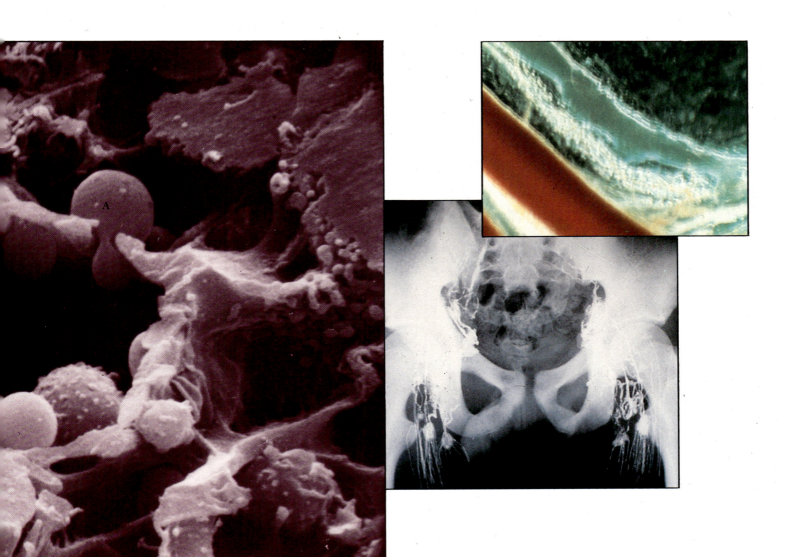

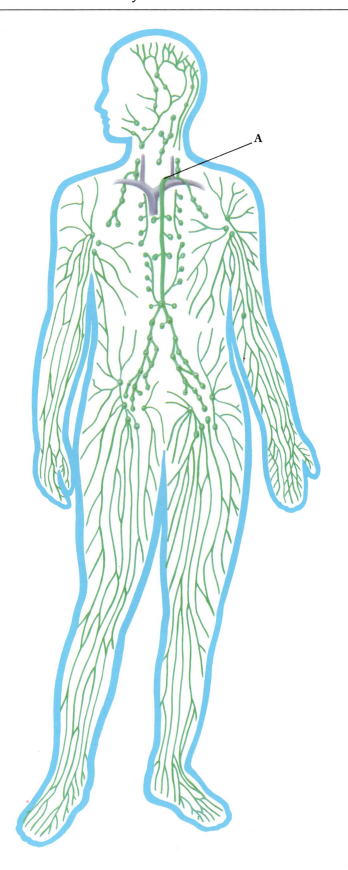

Destruction of invaders

The antibody molecules that we produce can protect us in a number of ways. As well as obstructing an individual molecule, an antibody can sometimes have the effect of sticking many molecules of toxin or virus together. This is possible because the antibody molecules are Y-shaped: two of the ends are able to recognize and stick to separate harmful molecules and, in this way, a whole chain or clump of them is held together and prevented from causing harm.

There is yet another way in which attacking bacteria can be rendered harmless. When the surface of a bacterium is covered with the appropriate antibody, this sets in motion a chemical destruction system in the blood which dissolves away the cell walls of the bacterium and kills it.

In fact, once the antibody clings to a particular hostile organism, it is a marked cell. In addition to the obstructive efforts of the antibody molecule itself, reinforcements arrive in the form of other specialized blood cells, macrophages, which are able to destroy whole bacteria by engulfing them and dissolving them away, and helping to neutralize poisons. It seems that the presence of the antibody indicates to many other cells that this particular organism is not part of the body but has come from outside and must be destroyed.

If our defence system turned upon our own cells with this ferocity, we would not survive for very long. And yet the sorts of molecules that stimulate antibody production are no different from the molecules that make up our own cells, and are assembled in similar ways. How is it that the body has lymphocytes that will only recognize intruders?

Scientists believe that in the womb and shortly after birth more types of lymphocytes are made than we need and that some of them may actually have the power to attack our own cells, to respond to 'self' as well as 'non-self'. In the thymus gland, where many of the lymphocytes are produced, all the cells that would be able to mount this sort of internal attack are identified and destroyed, leaving us with only those cells that attack 'non-self'.

Immunity

In the months after birth, while his own internal defence system is establishing itself, a baby is protected by some of the mother's antibodies which remain in his circula-

As the map shows, we have lymph vessels in every part of our body. At intervals they feed into strategically placed nodes, which act as filters and lymphocyte production centres. Eventually most of them join up at **A** and empty into the subclavian vein on the left side of the body; a much smaller duct empties into the right subclavian vein.

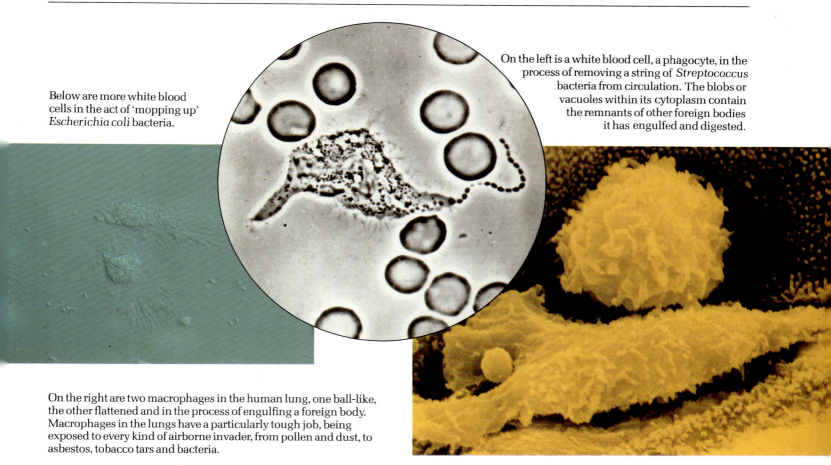

Below are more white blood cells in the act of 'mopping up' *Escherichia coli* bacteria.

On the left is a white blood cell, a phagocyte, in the process of removing a string of *Streptococcus* bacteria from circulation. The blobs or vacuoles within its cytoplasm contain the remnants of other foreign bodies it has engulfed and digested.

On the right are two macrophages in the human lung, one ball-like, the other flattened and in the process of engulfing a foreign body. Macrophages in the lungs have a particularly tough job, being exposed to every kind of airborne invader, from pollen and dust, to asbestos, tobacco tars and bacteria.

tion. If the mother has been exposed over the years to a range of diseases, she will have in her blood the defending antibodies to all of them. Her own cells will also have a 'memory' for diseases she has had in the past but she cannot pass this memory on to her baby.

We have seen how, when first infected with viruses or bacteria, the appropriate lymphocyte starts to produce daughter cells that can make large quantities of antibody. At the same time, the lymphocyte produces another type of offspring, a 'memory' cell, which will stay in the circulation until the same type of organism attacks again. This cell is similar to the original lymphocyte that first recognized the invader but because there are many more of them after an infection, the same organism will be recognized and attacked much more quickly next time. The presence of these memory cells means that an infection which we already have does not have time to take hold before it is destroyed: we are immune.

The baby, with no memory cells, has to start from the beginning with each new infection, unless he has some help in the form of immunizations. Because the lymphocytes respond to just a *component* of the virus or bacteria, that component can be injected into the tissues so that the immune system will produce antibody and memory cells for it in the same way as for a natural infection. Since the child is now equipped with supplies of antibody *and* with memory cells for that particular disease, he will be protected when he encounters it.

In the same way, someone with a cold will acquire immunity to that particular virus. Several days after the virus first attacks, the inflammation subsides, the virus is all but destroyed by the large amounts of antibody circulating in the blood, and the memory cells are on the lookout for the next entry of the virus.

But, as we all know, someone who has had one cold is *not* free thereafter from colds for the rest of his life. In the course of its evolution, the virus that causes colds has acquired a selection of alternative coatings, and is continuing to evolve. Since it is the virus's coat that stimulates the antibody cells, if the virus attacks wearing a different coat from last time, it will not trigger off memory cells and the symptoms will be just as bad. It is believed that the cold virus has a choice of 90 or so coats, so we can have a cold or two every year before we have acquired memory cells for every single one, and new ones are evolving all the time. In the ceaseless struggles between nature and the ingenious body, nature occasionally wins.

21 · Half Shares in the Future

The ability to reproduce develops fairly late in humans. Unlike many animals, we have to wait one fifth of our life-span before we can produce offspring. But that waiting time is not wasted: taking place under the surface of the growing bodies of children is a series of events that lead up to the most important act a human body has evolved to perform. Everything else we do – eating, drinking, breathing, moving – keeps us healthy enough to reach the age when we can reproduce. As individuals, we survive to reproduce. As a species, we reproduce to survive.

Although the differences are more apparent than the similarities, there are certain aspects of the reproductive system that are common to both sexes. Both have small amounts of tissue which produce the cells that will eventually combine to form the next generation – the ovaries in women and the testes in men. And they both have a delivery system that brings the cells to an accessible meeting place. One way in which the female differs, however, is that she has to house the baby during its development and so she has a womb where the fertilized egg can stay for nine months. With a contribution from both sexes, an ovum from the woman and a sperm from the man, the baby that results will have characteristics of both the parents.

At about the age of 11 years – slightly earlier in girls than in boys – the body goes through its most intense period of preparation for reproduction. Boys and girls start to make the transition to men and women. Even at this age they have usually acquired a certainty about their own sex and how they should behave. From now onwards this becomes gradually stronger, as physical events within their bodies lead them to follow behaviour patterns that have evolved for each sex to ensure that they make the most of their opportunities to reproduce. For the rest of their adult lives the male and female will each play a role which is a complex inter-action of their physical and psychological make-up.

Development of male and female foetuses

Many of the developments inside a child at puberty are programmed before birth, in the womb. Until six or seven weeks after conception, male and female foetuses look identical – they possess rudimentary sex organs and two sets of ducts (male and female). At that stage one of two things can happen: the foetus can continue along a predetermined path to develop into a girl, or this line of development can be prevented, and the foetus can turn into a boy.

Being female is basic to the developing child – only the presence of the Y chromosome in each cell, a contribution from the father's sperm, will channel the foetus towards maleness. If that chromosome is absent, the foetus develops the normal range of female sex organs and the male ducts disappear. By the fifth month in the womb, she has ovaries containing all the eggs she will ever have, and a womb and a vagina.

If the Y chromosome is present, however, the tissue that would have become ovaries turns into testes and the female ducts disappear. From this point onwards, the development of the other sex organs is under the control of the testes. They secrete a hormone that sculpts the familiar internal and external shapes of the male. Without this hormone, the organs would develop as female.

Once the sex is determined, the foetus starts to grow either testes or ovaries which will provide the germ cells – the ova and the sperm. A baby girl starts life in the womb with about seven million ova – an over-abundance to compensate for those that will be lost. At birth some degenerate, leaving a million or so. By the age of puberty at 11 or 12, she has about 300 000 ova left in her ovaries, five per cent of the number she started with. Each month after puberty some of them will be released, ready to be fertilized by sperm.

Like all body cells, the ova and sperm are produced by cell division. But they are different from all our other cells in one very important respect. Skin, bone, liver, muscle – all of them can be traced back to the earliest cells in the embryo containing 46 packets of information, the chromosomes, that specify exactly the design of the living body. Each time one of these cells divides, the nuclei of both cells still contain 46 chromosomes to pass on to the next generation of cells.

But because a new being is going to be formed from the merging of a sperm and an ovum, if each had 46 chromosomes there would be a surplus of information when they joined. Sperm and ovum cells therefore have 23 chromosomes each, formed from special ova precursor cells by a reduction division. In a girl foetus, it takes an unusually long time for them to complete this reduction process. Ova start to divide while the foetus is

At puberty the sexual identity implicit in our chromosomes becomes fully expressed – we develop 'secondary' sex characteristics. The girls in this picture will reach the prime of their reproductive lives at around the age of 24; the boys will reach it a little earlier, between the age of 20 and 24.

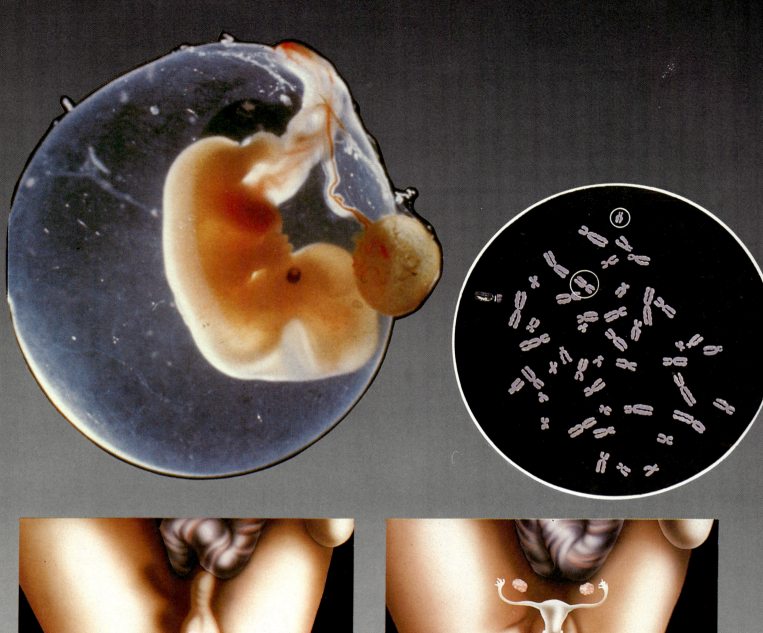

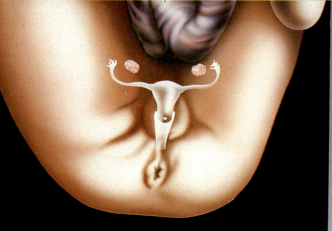

The developing embryo at the top of the page is six weeks old, and to all outward appearances is neither male nor female. But tucked away in every nucleus in every cell of its 15mm- (⅗in-) long body are 46 chromosomes, two of which hold the key to sexual identity. In the circular picture above the sex chromosomes – in this case an X and a Y – are ringed. An embryo with the XY combination will, at the age of seven weeks or so, begin to develop into a boy. Up until that time all foetuses look identical – the genital area is just a protuberance within a slit, with a swelling on either side.

In a boy the protuberance enlarges into a penis, the slit fuses, and the swellings develop into a scrotum. The result is shown above left – a male foetus at about 12 weeks.

In a girl the protuberance remains as the clitoris; the slit and the swellings become the vulva, with its inner and outer lips. A female foetus is shown above. Up until the seventh week foetuses of both sexes possess early traces of ducts that eventually become the vagina and uterus, and the vasa deferentia or sperm ducts. Then the appropriate duct develops and the other regresses. The germ cells that become the testes and ovaries do not start life in the genital region, but migrate into it after the seventh week.

still in the womb, but then stop dividing before they have finished, entering a state of suspended animation that could last as long as 50 years. Meanwhile they wait in the ovary for the moment of ovulation.

In contrast to girls, boys do not carry around a lifetime's supply of sperm cells. Perhaps this is just as well, because in an active life a man could produce enough sperm to populate the earth many times over if each was fortunate enough to meet an ovum. The male produces his cells almost on demand. When the boy reaches puberty, his sperm will start to be manufactured at a fairly steady rate, but production will only increase significantly if he becomes sexually active.

In the early years of the developing child, the sex organs are largely inactive. While sexuality may be part of the child's *mental* life, the physical possibilities are limited until a dozen or so years after birth. Then in a comparatively short time the young bodies experience far-reaching changes that turn children into adults and sometimes turn adults into nervous wrecks.

Puberty

The long period before the human animal is fertile is a preparation for adult life: it provides opportunities for the young to learn by imitating adults and by using their developing brains to think out things for themselves. This means that they only acquire a fully functional reproductive system after they have had enough time to learn how to protect the next generation during *its* formative period.

The onset of puberty brings a series of mental and physical changes that drive male and female to seek each other, and make it possible for them to produce new human beings. Their bodies are becoming equipped with the necessary physical characteristics and this results not only in the physical *ability* to have sex and produce babies, but also in the *desire* to do so, or at least to make love.

Both the physical characteristics and the mental and emotional processes depend on the coordinated release of hormones from different parts of the body, principally the brain and sex organs. The sequence of events centres on the brain, in particular the hypothalamus and the pituitary. During childhood, there is a gradual rise in the brain's output of two hormones called follicle stimulating hormone (FSH) and luteinizing hormone (LH), which cause a slow and steady increase in the size of the testes and the ovaries. The hormones are the same in both sexes but they lead to different events in males and females. Their effect is to stimulate the testes and the ovaries to secrete their own hormones, which spread throughout the body and start to produce the familiar signs of puberty.

The bodily changes that occur at puberty have several useful results. Primarily, of course, they prepare the ground for conception and fertilization. They also may help to make males and females more attractive to each other and increase their chances of reproducing.

Effects of puberty in girls

Puberty starts earlier in girls than in boys, although there is a wide variation between individuals in the actual age at which it begins. With girls, it seems to be closely connected with the age at which their mothers reached puberty. When a girl is about ten years old, the ovaries start to increase approximately tenfold in weight, under the influence of FSH and LH. It is this increase that leads to the secretion from the ovaries of the hormone oestrogen, which enters the bloodstream and is carried to all the body's tissues. Some tissues contain cells which are programmed to respond to oestrogen, in various different ways. Cells in the chest wall, for example, respond by growing and organizing themselves into breasts with glands that will be able to produce milk at the appropriate time. The bones that contribute to the height of the girl respond by suddenly growing at a much faster rate. Hair-producing cells become active under the armpits and around the groin. Internally, the uterus, which has been a tiny blob of inactive muscle, starts to increase in weight. In pregnancy it will become 25 times as heavy and able to carry a full-sized baby.

Perhaps the most conclusive sign that a girl is becoming a woman is the onset of menstruation. The menstrual period is actually the end of a four-week cycle of activity during which the womb has prepared for the task of housing and nourishing a fertilized egg, although conventionally menstruation is portrayed as beginning the cycle. However, a girl may not be able to conceive for two or three years after her first period, since her first egg is often not released until some time after menstruation begins.

Every 28 days or so, the female body undergoes an intricate sequence of changes with two possible outcomes: conception or menstruation. Menstruation happens about 400 times in every woman's life. The onset of the 28-day cycles may be irregular at first but, for reasons that are not clear, they become well established in most girls in a matter of months. There is some evidence that the cycle can change to fit in with events in the environment. Women who spend a lot of time together, for example, will often find that their menstrual periods become synchronized. This could work through the unconscious detection of a chemical scent secreted by women at certain times in their cycle.

Whatever controls the timing of the cycle, it begins with the release of FSH from the pituitary, a gland at the base of the brain. When FSH reaches the ovaries, it stimulates one egg follicle (or occasionally more) and rouses the ovum within from its state of suspended animation, so that it can complete its division. The follicle starts to develop a thick membrane and the ovum grows in preparation for leaving the ovary and passing into the womb.

About two weeks into the cycle, a surge of LH is produced by the pituitary. This causes the follicle to rupture and send the ovum into the abdominal cavity. Some women can actually feel this happening, as a short sharp pain in the lower abdomen. The feathery fingers of the end of one Fallopian tube pick up the egg and transport it down towards the uterus. While the follicle was developing, it produced high levels of oestrogen and progesterone which caused the lining of the uterus to grow in thickness, preparing the womb for the egg's arrival. The empty follicle persists and continues to secrete these two hormones.

By the time the egg arrives at the womb, having travelled down the Fallopian tube, a richly nourished womb lining awaits it, with a complex supply network of spiral arteries which keep the womb lining supplied with oxygen contained in the blood. Unless the egg becomes fertilized, in which case it starts to secrete extra hormones, the corpus luteum will stop releasing oestrogen and progesterone. As a result the womb will no longer be able to maintain its thick lining, which will start to shrink, constricting the blood vessels and making it more difficult for them to deliver oxygen. Starved of oxygen, cells in the lining of the uterus start to die. The blood vessels themselves degenerate and their walls rupture, releasing blood which washes away the remnants of the lining, producing the familiar signs of a menstrual period.

This system allows the body, whose whole 'purpose' is to reproduce itself, to make the most of every opportunity to do so. If a released egg is fertilized, the womb is in a good state of preparation for the growing baby. If fertilization does not take place, then there is a speedy return to the beginning of the cycle in case another opportunity arises.

Effects of puberty in boys

We saw earlier how many more eggs were produced in a girl than could ever be fertilized. This over-production, by which the girl baby starts with perhaps a million cells that could become ova, is positively frugal compared with the production of male cells. From the first stirrings of puberty in the teenage boy, his body is capable of delivering sperm cells at the rate of 200 million every 24 hours, and can continue to do so until old age.

Puberty for a boy, as for a girl, involves a range of bodily changes which alter his appearance and physical abilities, and which may make him more attractive or more competitive, and which ensure that he is able to produce and implant his sperm. His puberty starts a couple of years later than a girl's at which time he puts on a tremendous growth spurt as his long bones

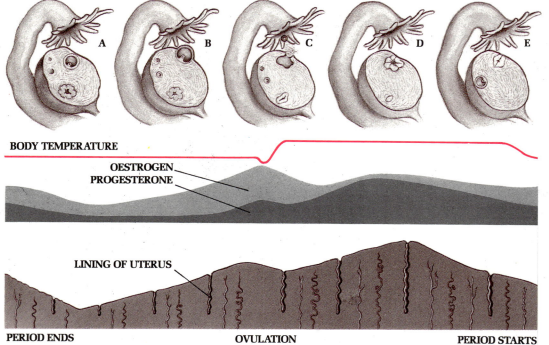

BODY TEMPERATURE

OESTROGEN
PROGESTERONE

LINING OF UTERUS

PERIOD ENDS   OVULATION   PERIOD STARTS

Each menstrual cycle starts with a ripening ovum contained in a follicle within an ovary **A**. As levels of the female hormones oestrogen and progesterone rise, the follicle swells **B** and then releases its ovum **C**. This is ovulation, marked by a peaking of oestrogen production and a distinct rise in body temperature. Already the lining of the uterus has become thick and blood-rich under the influence of progesterone. The empty follicle then turns into a miniature gland, the corpus luteum **D**, which secretes more progesterone in anticipation of a fertilized egg implanting itself in the uterus wall. If this fails to happen, preparation for it tails off **E**, and the lining of the uterus breaks down and is lost – a period.

The pea-sized pituitary gland which hangs from the underside of the brain, as shown below, is the orchestrator of femininity and fertility. By sending out chemical messages in the form of follicle-stimulating hormone (FSH) and luteinizing hormone (LH) it initiates and coordinates the events of puberty and after that the monthly cycle of menstruation. It also manufactures a variety of other hormones that influence general growth and metabolic rate, and specifically the thyroid glands, the adrenal glands, the kidneys and the milk glands in the breasts.

From the age of 10 onwards a gradual rise in blood levels of FSH and LH affects many organs, but in particular the ovaries (below right) which grow and start to produce hormones of their own.

The changes of puberty - breast enlargement, laying down of fat, growth of pubic and underarm hair, and so on - are brought about by ovarian oestrogen. The change from girl to woman is signalled by menarche, the first menstrual period. After that, the monthly ripening of an ovarian follicle is stimulated by a rise in FSH output from the pituitary. A surge of LH two weeks later causes ovulation.

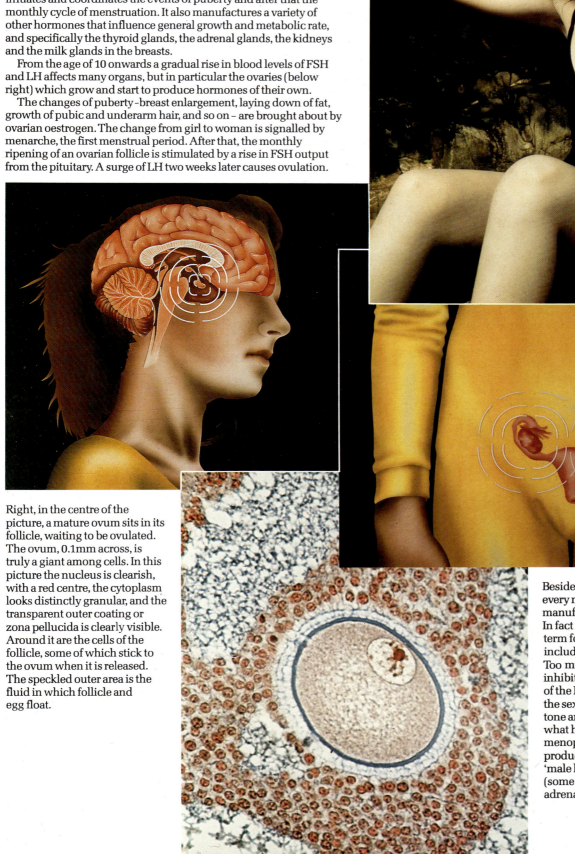

Right, in the centre of the picture, a mature ovum sits in its follicle, waiting to be ovulated. The ovum, 0.1mm across, is truly a giant among cells. In this picture the nucleus is clearish, with a red centre, the cytoplasm looks distinctly granular, and the transparent outer coating or zona pellucida is clearly visible. Around it are the cells of the follicle, some of which stick to the ovum when it is released. The speckled outer area is the fluid in which follicle and egg float.

Besides producing a ripe ovum every month, the ovaries manufacture oestrogen. In fact 'oestrogen' is a blanket term for several active substances, including oestradiol and oestrone. Too much at the wrong time can inhibit ovulation (the principle of the Pill) and too little causes the sexual tissues to decrease in tone and robustness (which is what happens after the menopause). The ovaries also produce small quantities of the 'male hormone' testosterone (some is also produced in the adrenal glands).

The growth spurt that boys undergo during puberty is seen in the line-up of boys above, aged from 11 to 17. The average age for maximum growth rate in European and North American boys is 14, with an average height gain of 9cm ($3\frac{1}{2}$in) in the fifteenth year and a gain of 7cm ($2\frac{3}{4}$in) in the fourteenth and seventeenth years. Girls reach the peak of their growth spurt two years earlier on average. Just before puberty both girls and boys are growing slower than at any time since birth.

Along with boys' increased muscle bulk and power comes an increase in competitive and combative drives. Meanwhile, inside the seminiferous tubules in the testes, sperm are being manufactured. The photograph on the opposite page shows sperm tightly packed inside a sectioned tubule. As they mature they move inwards – the tangle of tails in the centre of the tubule are those of mature sperm awaiting ejaculation.

The hormone that brings about these changes is testosterone, shown on the far right in its crystalline form.

increase in length. A younger girl might have been taller than him for a brief period while she was starting puberty, but now he will shoot up and surpass her, and remain permanently taller as they both continue to grow. His skin and hair undergo important and not always pleasant changes. Pubic hair grows, in a pattern that is distinctly different from the female triangle, and it is followed by hair under the arms and on the face. Unfortunately, this hectic overactivity of skin and hair cells can lead to increased oiliness of the skin, and acne. The sex organs themselves, penis and testes, steadily increase in size and, as part of the general increase in virility, the male voice deepens as the larynx grows and the vocal cords double in length.

All these changes are triggered by the hormone testosterone that is now being produced by the testes. (Again, FSH and LH are responsible for stimulating this hormonal production.) In the testis itself are hundreds of metres (yards) of tubes whose walls are the breeding place for sperm. A single tubule is 0.5 metre (20in) long, and its walls are lined with cells that will give rise to wave after wave of sperm cells. Like the ovum, the mature sperm has only 23 chromosomes, and it achieves this by the same reduction division, in the lining of the tubule – a process which takes 70 days. During that time millions of sperm cells are produced, starting as round immature sperm and ending up as torpedo-like devices consisting of a dense head, packed with the father's information in the form of 23 chromosomes, and a long active tail to propel the sperm to its target.

One of the many interesting differences between men and women is the fact that women produce their egg cells *inside* the body while men, to all intents and purposes, produce theirs *outside*. While the scrotum is of course attached to the body, it is surrounded on most sides by the outside world rather than the cosy warmth of the body's interior. It is not a very safe place for such important organs to be, so some other factor must have been more important than vulnerability when our organs evolved. That factor turns out to be temperature. At 37°C (98.4°F), the body's normal temperature, production of sperm almost ceases. It could be that, to make the process as efficient as possible, an intrusion into the cooler outside world had to be developed, so that the 'sperm factory' was consistently cooler than the rest of the body.

Some of this coolness comes from the circulation of air around the scrotum, and anything which prevents that can slow down sperm production. There is even a heat exchange system so that the arteries bringing blood to the testes are cooled by the veins returning to the body. The development of tight trousers for men could well herald the downfall of the human race, or at least of the more fashion-conscious members. Similarly, regular hot baths can reduce the sperm count below the level at which fertilization can take place, although to achieve efficient contraception by this method it would probably be necessary for a man to spend most of his time in the bath.

With millions of mature sperm waiting in the testes, and the ovum waiting in the Fallopian tube, the stage will be set every month from puberty onwards for the meeting of the two.

22 · Coming Together

Sex is a powerful force, and society constantly reminds us that women were made for men and men for women. Whether we like it or not, sexual reproduction is the ultimate purpose of the living body: each of us can be seen as just a device for ensuring the survival of the species by passing on some of our characteristics to the next generation, although evolution sees to it that we can enjoy ourselves while we are doing it.

At the core of the physical and emotional events of sex lies one simple purpose – to mix the characteristics of two people to make a third person. This is the way our species adapts or improves itself. Nature has devised a system for making sure that individuals with new talents are continually being created: sexual reproduction results in unique new humans because, in order to produce a child, a man and a woman – each of whom have their own individual qualities – have to create the meeting of two cells, one from each of them.

Virtually all cells of the body have 46 chromosomes each, which provide the genetic code for all our characteristics. The only exceptions are ova and sperm, which each have 23 chromosomes (and red blood cells, which have no nucleus). On joining at fertilization, the full complement of 46 chromosomes is restored – 23 *pairs*. One of each pair comes from each parent and codes for the same characteristics, but for variations of them. For example, a child inherits two sets of genes coding for eye colour which are found on the same pair of chromosomes, but the father's chromosome may carry genes for brown eyes and the mother's for blue. In this way features of the child will resemble one parent or the other, depending on which of the genes in the mixture are dominant and are displayed.

In preparation for fertilization, the consignments of 23 chromosomes are packaged differently according to the sexes. The woman's contribution, the ovum, is large in relation to other cells as it contains stores of nutrients, enzymes and proteins which will help the egg to develop once it has been fertilized. The sperm of the man, however, is much smaller, with a long, whiplash tail to enable it to move towards the ovum in the uterus of the woman.

The events that lead to conception begin with the attraction that each parent feels for the other. The urge for sex is usually stronger than the more considered desire to produce children. There seem to be two components that lead males and females to a sexual relationship. First, there is often a strong drive to seek affection. This leads to the mysterious phenomenon of love, where a particular individual captures the attention of another, leading to a desire to be close to him or her. The roots of this attraction are unfathomable. Some of it may be innate but it may also have a lot to do with the early years of childhood and the development of our individual understanding of human relationships. Secondly, there is a strong pleasure to be found in the act of sexual intercourse itself. There are powerful mechanisms in the brain that participate in the sex act and lead to a desire to repeat it. These two factors lead man to seek sex. Most other animals, in contrast, restrict sex to a seasonal phase, the time when they are able to conceive and feel the urge for sex, interspersed with times when they cannot conceive and have no sex.

Our development, from humans who behaved like apes, only receptive to sex at certain times, into humans who exhibited much greater cooperation between the sexes (because of their greater sexual availability) has had advantages for the children. The women who were

Below, a sperm penetrates an ovum and the 23 chromosomes packed inside its head unite with the 23 chromosomes of the ovum. Its tail, having propelled it to its goal, breaks off.

The photograph shows a fertilized ovum, its granular cell body surrounded by several jelly coats; these thicken and harden to prevent other sperm entering.

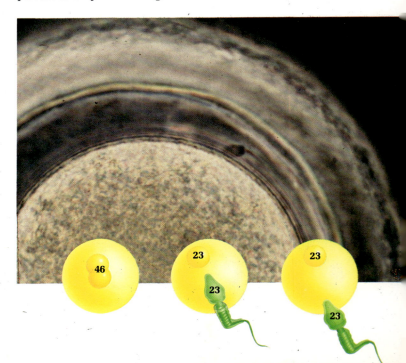

The force of attraction between male and female makes the world go round, and the process of sex inheritance, shown in the diagram below, ensures that there are roughly the same number of men and women in the population.

In each of her body cells a woman has 44 chromosomes that have nothing to do with sex plus two more, both Xs, which make her female. A man also has 44 chromosomes that have nothing to do with sex plus two more, an X and a Y, which make him male. During reproduction only half of our chromosomes – 23 – go into our sperm or ova, so that when a sperm fuses with an ovum the number of chromosomes reverts to 46.

It follows that each egg or sperm can only contain one sex chromosome. In an egg, the sex chromosome can be an X or an X! In a sperm, it has to be an X or a Y. Whether it is a Y-carrying sperm or an X-carrying sperm that succeeds in penetrating an egg is a matter of pure chance. If X-carrying, the child will be a girl, if Y-carrying a boy.

Attempts have been made to separate X-carrying from Y-carrying sperm, even to disable one rather than the other, but so far they have been unsuccessful.

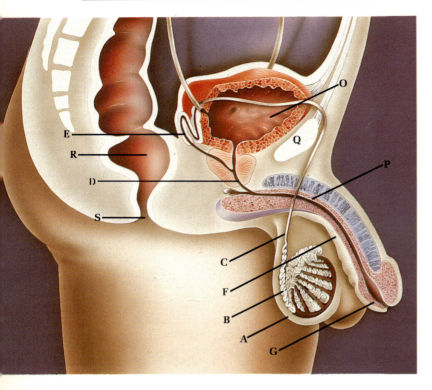

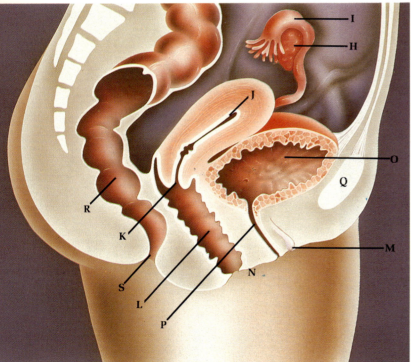

The drawings above show the main differences between the male and female urinary and genital organs. In the man **A** is the testis, **B** the epididymis, **C** the vas deferens, **D** the prostate gland, **E** the seminal vesicle, **F** the penis and **G** the glans penis; in the woman **H** is the ovary, **I** the Fallopian tube or oviduct, **J** the uterus, **K** the cervix, **L** the vagina, **M** the clitoris, and **N** the vulva. Then, in both sexes, **O** is the bladder, **P** the urethra, **Q** the pubic bone, **R** the rectum and **S** the anus.

available all the time would probably be cared for more, eat better food and be protected, together with their children, from predators or aggressors. Perhaps more importantly, the constant availability of the female, combined with her desirability to the male, maximized the chances that they would come together to reproduce and cooperate in caring for their offspring.

Arousal and intercourse

It may take some time for an initial attraction between two people to turn into a desire for sex, and the next series of events can happen as quickly or as slowly as both of them choose. Our bodies are designed to unite a healthy sperm with a healthy egg. The shape and location of the sexual organs and the physical events of intercourse have evolved to make each stage of the activity as efficient as possible and to increase the chances of a successful fertilization taking place. But the key factor for the physical events of sex is usually a number of psychological events.

Sensory perception is often the starting point, and plays an important part. If the skin is touched around the sex organs, or elsewhere on the body, a whole variety of reactions can occur. Like any touch signal, the sensory message travels from the point of contact to the

As soon as a mature egg is released from the ovary it is wafted into the delicate, feathery funnel at the upper end of the Fallopian tube, shown on the right. The egg is then ferried along the tube by the tiny, waving hairs that line its walls – these are shown below, magnified 3000 times. Tufts of longish hairs are irregularly interspersed with cushions of microvilli, projections tinier still.

spinal cord. There it can trigger a response in the sex organs, such as swelling and increased secretion of lubricating mucus.

Simultaneously, however, the sensory message continues up the spinal cord to the brain. There the sexual response becomes conscious and can trigger all sorts of sensations that may enhance or diminish the experience. If the touch comes from an unfamiliar source or is unwelcome, then brain activity can inhibit the sexual response. If the person who causes the sensation is known and loved, the physical response that has started at the level of the spinal cord is likely to continue, and it will gain renewed strength if it is accompanied by appropriate signals from other senses, such as vision, smell, hearing and even taste.

At this stage, the body is using signals of several different sorts to build up to an appropriate sexual climax. Our response to some of the sensory signals may be innate – a matter of instinct. Being touched, the sight of the sexual readiness of the opposite sex, and the smells of secretions that occur during sexual excitement may well be experiences we do not have to *learn* to be aroused by – it may just be human nature.

Our response to other signals may be learnt. Different human societies find different stimuli arousing: what is attractive to one group of humans may not have the slightest effect on another. And even within the same society the sexual tastes of individuals may vary widely. But nearer the heights of sexual arousal, the similarities become more apparent, and the events which lead to the peaks of fertilization and conception are the same in everyone.

Every element in sexual activity can be seen as an attempt to smooth the passage for the sperm and the ovum and bring them as close to each other as possible. In the man, erection of the penis permits its insertion into the woman's vagina and shortens the sperm's journey to the depths of the Fallopian tubes. Normally most of the blood flow around the male sex organs bypasses the penis, but during sexual arousal the arteries leading to those organs dilate and blood flows into spongy areas of tissue that expand to produce erection. At first, the blood flowing in is much more than that flowing out and so the organ increases in size. Then the veins dilate to balance the input, but still remain narrow enough to create a back-pressure that retains the erection. In the woman there is a similar swelling of the genital area which helps to heighten the sensations of intercourse; in this way, the movements that lead up to ejaculation can cause increased

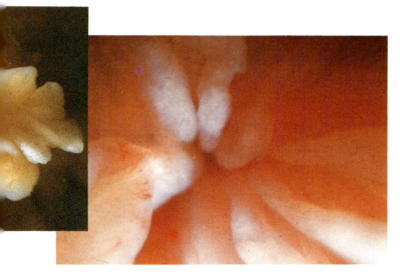

Above is an egg's eye view of the narrow entrance to the Fallopian tube. The extended fingers or fimbriae that lead into the tube are positioned very close to the ovary, ready to catch eggs as they are released and guide them into the tube. Rarely the fimbriae fail to capture an egg and it floats free in the abdominal cavity; if it is fertilized here, by a sperm that manages to swim all the way up and out of the Fallopian tube, it may develop and even implant in an abdominal organ. This is termed an abdominal ectopic pregnancy; but the embryo soon gets into difficulty and aborts.

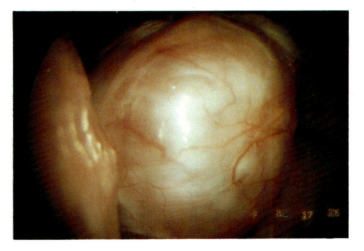

Above is a surgeon's eye view of an ovary, seen through a special instrument called a laparoscope – the large spherical body, with tiny blood vessels on its surface, is the ovary. The white blister in the middle is a Graafian follicle, containing a ripe ovum floating in fluid. It is possible to collect ripe eggs, for *in vitro* fertilization for example, by inserting a thin tube into the abdominal cavity along with the laparoscope; the tube sucks up the pinhead-sized egg as it floats free of the Graafian follicle.

secretions in the vagina so that when the sperm eventually emerge they find a moist and receptive environment in which to continue their journey.

The back and forth movements during intercourse produce orgasm, resulting in muscular contractions in the tube that leads from the testes to the penis. Once these contractions have started they cannot be stopped: inexorably, about 100 million sperm travel by a roundabout route from the testes to the penis, along a tube that leads upwards to the level of the bladder, inwards to go around the bladder and meet up with another tube carrying secretions from the prostate gland which help to smooth the passage of the sperm on the rest of its journey outwards along the penis. Although the sperm in the testes are only a few centimetres from the tip of the penis, they have to travel about 40 centimetres (16in) before they are ejaculated.

Journey of the sperm
When the sperm emerge they are deposited near the cervix – the neck of the womb. This is really the beginning of the journey to meet the ovum. From now on the sperm is on its own and has to rely on the propulsive action of its long tail, occasionally helped or hindered by the activity of the womb and the Fallopian tubes, and the chemical climate it encounters there.

For the outcome to be successful, there has, of course, to be an ovum waiting. It may even be that the heightened activity of the woman's body during sexual intercourse sometimes brings on ovulation, which would certainly improve the chances of fertilization.

The open end of the Fallopian tube waits near the ovary to capture the egg when it is released. The egg is then pushed towards the womb on a current of mucus by the movement of fronds around the side of the tube.

When the sperm have been deposited in the vagina, they coagulate together for 20 minutes or so, a convenient method of stopping them slipping out again. Then the next stage begins. From the neck of the womb, the individual sperm cells still face a daunting journey. At each stage there is massive wastage of sperm and large numbers fall by the wayside on the journey towards the ovum. Some may not even find their way to the neck of the womb. Those that do, swimming at a speed of three millimetres an hour, may find the opening blocked by a plug of mucus that only dissolves around the time of ovulation. Many of the sperm fail even to get through this first barrier.

Those that are close enough to the opening of the womb at the right time of the month can swim through to the womb. They may possibly be helped by some of the muscular activity that accompanies intercourse: there is some evidence that at the time the woman reaches her climax, pressure drops in the womb and this helps to suck the sperm through the neck of the womb and onwards towards fertilization. Once inside the womb, the sperm are carried on gentle waves of muscular movement to the top end, where the two Fallopian tubes have their openings. This is where fertilization takes place – the uniting of sperm with egg.

But each month, only one of the Fallopian tubes contains an ovum, under normal circumstances, and

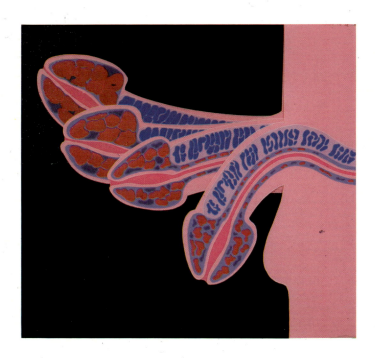

The two tubes that carry sperm from the testes to the penis are the vasa deferentia, one of which is seen in cross-section above. The thick wall is mostly muscle tissue that contracts in peristaltic waves to propel sperm and fluid through the tiny duct in the centre.

During arousal (left) the penis changes from small and flaccid to large and rigid, and capable of penetrating the vagina. Blood accumulates under pressure in the corpora cavernosa (blue) along the upper part of the penis, and in the glans (red) at its tip.

Biologically speaking the coming together of a woman and a man has one main purpose: the coming together of an egg and a sperm. During sexual arousal and foreplay blood flow round the body changes, to supply and swell the sexually-sensitive areas. Some of these can be seen in the thermographic picture on the right, as white and red 'hot spots' – mouth, cheeks, neck and chest.

Although a couple shares love, the size of the cells they contribute (shown below) is definitely unequal. The woman contributes a large cell by cell standards – 0.1mm in diameter and just visible to the naked eye. The man contributes a cell that is 200 times smaller, but what sperm lack in size they make up for in numbers; some 100 million of them are ejaculated at a time. Only one fertilizes the ovum.

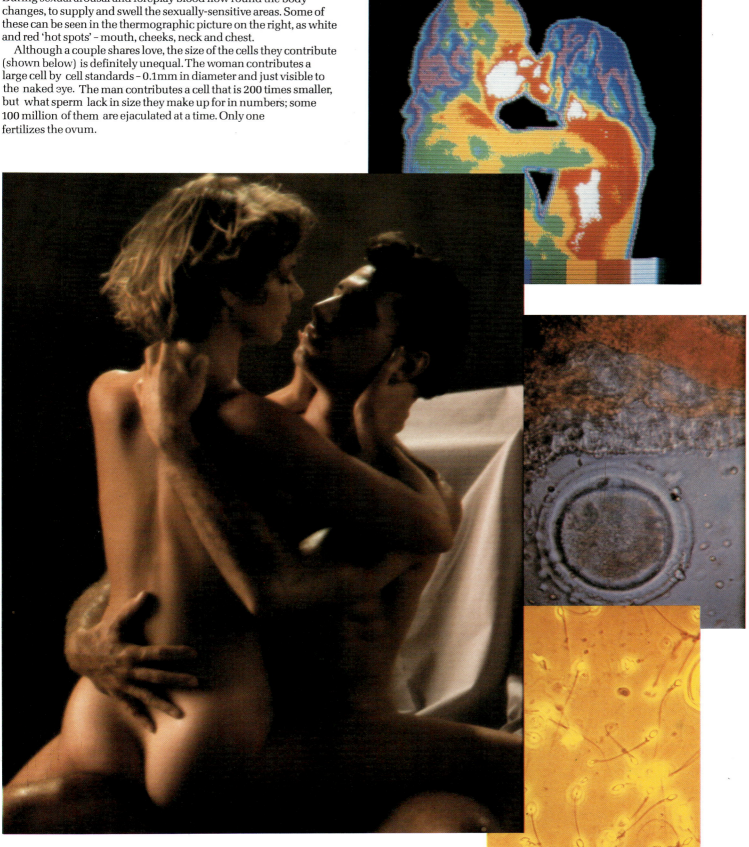

about half the sperm will enter the wrong tube. This means that only a few hundred sperm – a fraction of a per cent of the original number – will have found their way to the tube with the egg in it. For those that enter the correct tube, it is a race to find the newly released ovum ; only one sperm out of the 100 million starters will be successful. It has taken the sperm about six hours to reach this point.

During its journey through the uterus there have been subtle changes in the sperm that prepare it for penetrating the ovum. One result is to arm it with enzymes which will dissolve the outer membrane of the egg, when it is found.

As the first sperm nears the ovum, substances emerge from the membrane of the egg and dissolve part of the head of the sperm. Then the enzymes are released from the sperm, which in turn help to penetrate a barrier surrounding the ovum. Although the ovum is entirely surrounded by sperm, only one will penetrate it. Some of the others are reduced to helping the winner make its way deeper into the egg towards the nucleus. At the moment of fertilization, the winning sperm buries itself inside the ovum and loses the powerful tail whose whiplash movement has helped to bring it all the way from the neck of the uterus. A new human being starts to be formed, with its full complement of 46 chromosomes.

Only now is one very important fact established – the sex of the child. In about half the sperm one of the 23 chromosomes was shaped like an X, and in the rest the corresponding chromosome was shaped like a Y. These special X and Y chromosomes carry information that refers specifically to the sex of the child. If it happens that the successful sperm has an X chromosome, the fertilized egg will develop into a girl, and if it has a Y chromosome it will be a boy. Thus it is the father who determines the sex of the children – the ovum always carries an X chromosome.

Whatever the sex, if all has gone well the beginnings of a baby are now in the Fallopian tube. It will be some time, however, before the parents-to-be know anything about it at all, although they are at least aware of the possibility that intercourse may lead to pregnancy.

It is likely that for early members of the human race, just as for animals, the acts of sex and birth were not linked at all. Why should the pleasurable activities of spring be connected in any way to the child who is born in winter? Humans and animals have a drive towards sex that does not depend on the decision to reproduce. In the interests of the continuation of the species, sex has evolved as a (usually) pleasurable way for people to unite their chromosomes to form a new individual.

Advantages of sexual reproduction

The real usefulness of sex is to be found in the *mixing* of two different sets of chromosomes. It is not necessary for reproduction to involve sex at all – it is possible to imagine a human race entirely consisting of women who gave birth at regular intervals to replicas of themselves, derived from cells in their bodies which contain identical genetic information to that contained in every other cell.

Some primitive living creatures do reproduce this way. But very early on in the evolution of life, the advantages of sexual reproduction became overwhelming. In a world in which the external environment is changing, a species which was unable

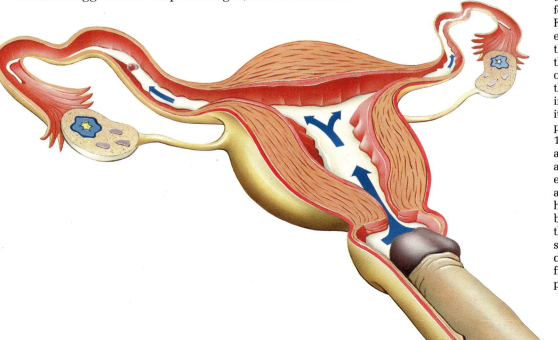

The formidable journey facing 100 million sperm is shown in this cut-away illustration of the female vagina, uterus and Fallopian tubes, with the ejaculating penis lying inside the vagina. All the way along the route (blue) sperm run out of energy and die, and half of them make the wrong turning, into the tube without the egg in it. In fact very few sperm – perhaps between 1000 and 100 000 – enter the correct tube, and of those perhaps only 100 actually reach the ovum. An enzyme released from the acrosomal cap enclosing the head of each sperm helps to break down the membrane of the egg. So although only one sperm penetrates the egg, the combined quantity of enzyme from many makes penetration possible.

to adapt would soon die out. A race of identical members could only survive in a stable world of constant temperature and steady unchanging food supplies, unworried by predators or competing species. The moment any of these factors changed in a direction that was disadvantageous for any one member, all members would suffer, being identical.

If half the characteristics of a new individual come from outside the mother – from another member of the species – the next generation will be different from its parents. In a large population characteristics would combine at random, and there would be an opportunity for all sorts of individuals to appear, with characteristics and abilities that had never existed in previous generations. In a species with a wide range of characteristics, there would be a greater likelihood that enough individuals would have the resistance or stamina or courage needed to survive whatever the harsh world threw at them and to ensure the survival of the species. These individuals then tend to produce offspring with some of the same characteristics.

White and black, yellow and brown, tall and short, fat and thin, the variation made possible by sex is a persuasive reminder of how, in yet another aspect, the functions of the human body have developed to become so well suited for the tasks it has to perform.

Most women are not aware that fertilization has taken place until some two weeks later, when they miss their period. Nevertheless, on average only one in two fertilized eggs lives long enough to implant itself in the uterus; the rest, presumably the least healthy and least likely to survive, regress or disintegrate.

Below is an egg photographed four or five hours after fertilization; it floats in fluid inside the Fallopian tube, surrounded by miscellaneous smaller cells and other debris. A few remaining sperm (three or four are visible in the clear area at top left of the picture) still swim hopefully round it.

23 · A New Life

One of the body's most remarkable abilities is the way in which it can make a copy of itself. Over nine months, a steady supply of basic ingredients can be assembled into a living, breathing, moving human being.

This all happens according to a detailed plan that is laid down in every cell of the developing baby's body. Almost like a living sculpture, a new human being is shaped from a mass of unorganized tissue.

From the moment of conception, the child-to-be has similar needs to those of a fully developed person. Regular supplies of food and drink, warmth, vitamins, immunity from disease, freedom from stress, exercise – all of these are provided during the first nine months of life. Even before a woman knows she is pregnant, the reproductive system of her body has carried out many of the most important steps towards providing her with a healthy baby.

Within a day or two of intercourse, her ovum, which is in one of her Fallopian tubes, has received one of her partner's sperm and incorporated his chromosomes into its nucleus. The fertilized egg now has a full complement of 46 chromosomes which, between them, contain all the information necessary to build and maintain a unique human being, a combination of the mother and father.

One of the first effects of fertilization is to release hormones that prevent the occurrence of the menstrual period that would have been due in a couple of weeks, and most women first learn that they might be pregnant when there is no loss of blood at their normal time of the month. Instead of shedding its lining, the womb – which is at this stage only about seven centimetres long – will have to prepare for pregnancy by thickening and enlarging itself many times.

Early development of the foetus

The fertilized egg is only a fraction of a millimetre across, but its size increases rapidly through repeated cell division, and by the time the menstrual period would have been due the developing child is likely to be made up of thousands of cells. Each time a cell divides it makes a copy of all its chromosomes for the new cell to take with it. One cell becomes two, two become four, four become eight, and so on, until the developing embryo contains millions and millions of cells, all with identical chromosomes. The embryo starts off as a symmetrical ball of cells that looks rather like a blackberry, called a morula, which rolls down the Fallopian tube towards the womb. It then implants itself in part of the wall of the womb and prepares to stay for the duration of the pregnancy. Only if implantation is successful, several days after fertilization, does pregnancy truly begin.

At the moment of implantation, the morula is made up of identical cells. But, of course, they will not remain identical. Although they all have the same chromosomes, they will eventually have to become different parts of the developing child. Within two weeks of fertilization there are the beginnings of primitive specialization, as some cells cluster in the middle of the 'blackberry' and others surround them. The ones in the middle will form the embryo while the surrounding cells will form the placenta, the embryo's life support system.

One of the biggest mysteries of biology is how the individual cells in a developing embryo know which part of the baby to form. This handful of cells will become a human being made up of 30 trillion cells, carrying out many different tasks as brain, skin, bones, heart and so on. And while they are developing into those particular tissues, they also have to know where in the growing body they should be located.

As the child in the womb develops, the ball of cells takes on a shape. Just as a pullover is knitted by following a pattern, a series of simple instructions carried out one by one leads to a complex end-product. Although a knitting pattern contains all the information needed to make the pullover, it would be very difficult to tell by looking at the written pattern what the end result would be like.

In the centre of each of our living cells is the pattern, or genetic code, for our bodies. The fertilized egg contains one copy of that pattern, written on 46 'pages', the chromosomes. As the foetus develops, the instructions in the chromosomes are followed step by step. Sometimes only a single 'page' is read and the rest of the pattern ignored. As the embryo develops, one cell might need to become a bone cell, for example, and will only need to follow the instructions that lead to bone formation. Another cell might develop into the beginnings of the nervous system, following a different section of the overall pattern. It is as if each cell has a

In the scan below, of a six-month-old foetus, a hand and an arm are plainly visible, and so are the two hemispheres of the brain.

Five and a half months after conception the foetus is about 30cm (12in) long and growing fast, but the proud mother's bulge is chiefly fluid in which the foetus floats surrounded by the amnion, a thin veil-like membrane. The foetus' eyes have now formed, and may even detect light, but the eyelids have grown over them and fused (they separate again later). The tiny fingers have fingernails, and the mouth opens and closes to swallow amniotic fluid, or occasionally to suck a thumb.

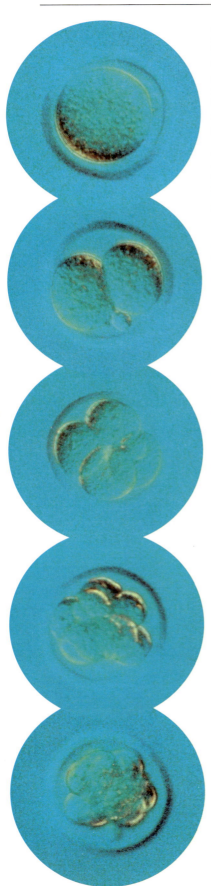

A new life begins with the series of divisions shown on the left. The fertilized ovum splits into two cells, then four, then eight, and so on. After three or four divisions the giant single cell that was the egg becomes a collection of more normal-sized cells.

On the right is an embryo at the morula stage, making its way through the much-folded lining of the Fallopian tube. The hairs that help it on its way are noticeably longer and more active immediately after ovulation, no doubt under the influence of oestrogen and progesterone, produced in considerable amounts at this time.

By the time it arrives in the uterus, about five days after fertilization, the morula has become a ball of a hundred cells or more, a blastocyst. And already there are obvious differences between the cells; the outer cells are distinctly flattened or cuboidal, and the inner ones are spherical.

A day later the blastocyst runs out of food reserves and buries itself in the nourishing lining of the uterus, below.

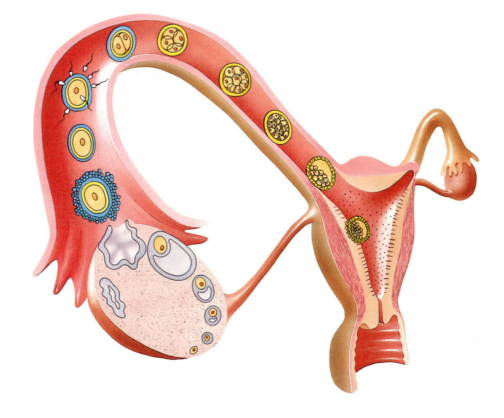

Thirty days after fertilization a human embryo, about 40mm (1½in) long, is little different from that of most other mammals. It is developing 'head first', from the head downwards. In this photograph the brain, and next to it the two red bulges of the early heart, are on the right. Curling upwards on the left is the tail. To the left of the heart is a dark bulge, the beginning of an arm. Half way along the tail is another dark area, the beginning of a leg. Throughout embryogenesis the arms develop a few days ahead of the legs.

During these first few weeks human embryos seem to recapitulate certain features of vertebrate evolution – at one stage they have gills, then lose them, and briefly they have a tail, then lose it.

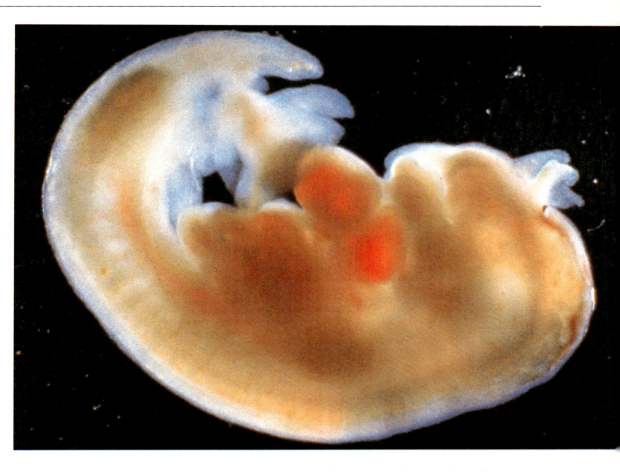

whole book of patterns for the whole body, but only ever follows one pattern for one part of the body.

Of course, the information in chromosomes cannot be seen, and can only be detected indirectly. Each chromosome is a coiled coil. Only when it is unwound can we begin to appreciate how much information is stored on each one, written in the genetic code. During development, sections of each chromosome are uncoiled and read in a long line, instruction by instruction. Each instruction leads to another element of the developing child being constructed out of the raw materials of the cells and the mother's body.

It is under this control that the early developments of the embryo take place. About three weeks after fertilization, after implanting itself in the wall of the womb, the cluster of cells pushes finger-like protrusions deeper into the wall, preparing to build new blood vessels that will link with the mother's own so that the rapidly growing child can become part of the mother's circulation and use her supplies of oxygen and food.

The cells of the embryo form a flat disc, which then starts to develop the bumps and dents that will later become recognizable features. One end becomes more pointed – this will be the baby's bottom – while the head end is more rounded. A dent becomes a groove and the

groove becomes a tube to form the gut. Initially just a straight tube, it develops kinks and expansions and finally becomes the digestive tract, from mouth to anus. By a month or so after fertilization, the embryo has the beginnings of recognizable features. A developing face, arm and leg buds, and, astonishingly, a beating heart.

One important characteristic that is impossible to distinguish at this stage is its sex. Although all of its cells have a sex chromosome that will make it a boy or a girl, this has not yet had any effect. Instead, the embryo has a ridge of genital tissue that could become either ovaries or testes. The size of this entire complex bundle of human cells at about one month is little larger than a grain of rice.

By about two months into pregnancy, the foetus is recognizably human. Though only 5-6 centimetres (2in) long, all the organs are formed, and the face has quite a human appearance. The head is the most prominent feature, taking up about half the distance between crown and rump. In the coming weeks it will grow more slowly than the rest of the body, resulting in the more recognizable proportions of a baby.

Although the tiny body is developing some of the signs of independent life, it will be many months before it will have enough of its systems working to be capable

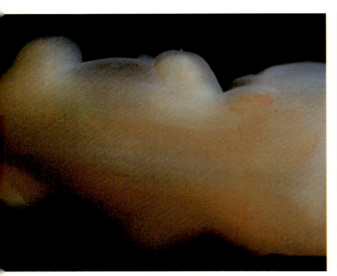
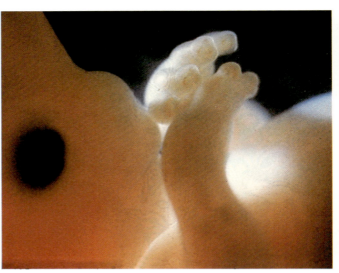

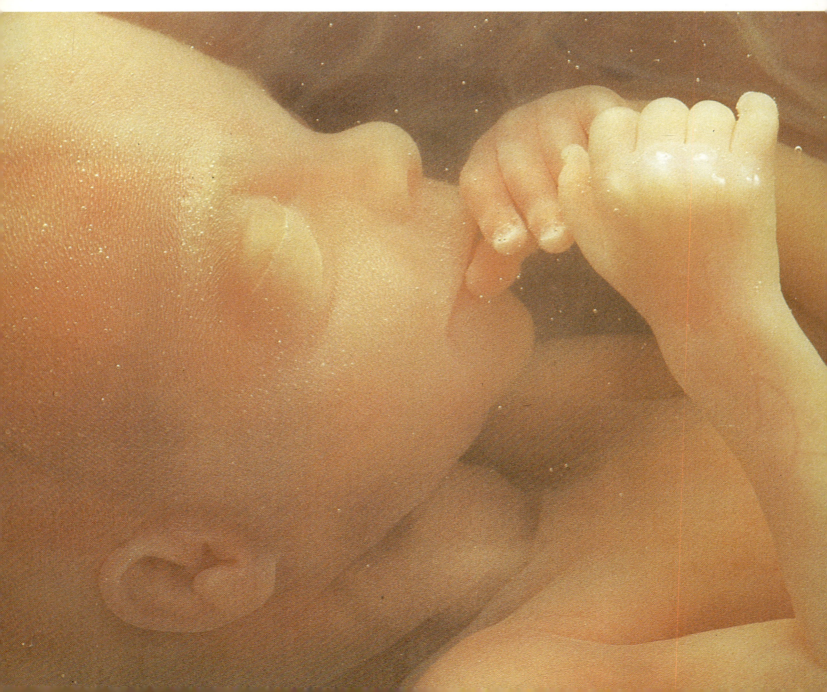

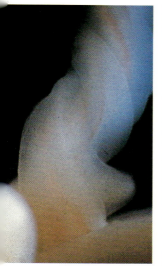
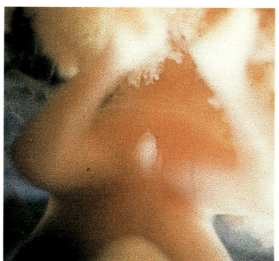
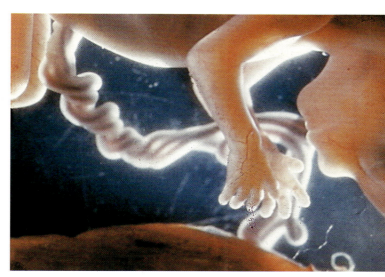

of surviving outside the womb. For the time being, it depends entirely on its immediate environment for every necessity of life.

While the mother concentrates her attention on the foetus, her own body is undergoing changes in almost every organ and system. Her blood volume increases by about 50 per cent, and the output of her heart goes up as well, by 30 to 50 per cent.

In the early stages of pregnancy the kidneys increase their output of water; this diminishes later in pregnancy, but towards the end the baby's head pushes against the bladder, and so the mother urinates more frequently again, but for a different reason.

The joints relax, particularly in the pelvis, and lead to changes in the way the pregnant woman walks.

Between them, mother and baby gain about 12 kilograms (26lb) in weight. The average loss at birth will be about 8 kilograms (17lb), leaving the mother 4 kilograms (9lb) heavier than she was at conception.

Many of these changes are controlled by two hormones, released partly by the mother's ovaries and partly by the tissues that accompany the developing foetus, the corpus luteum (the remains of the egg follicle in the ovary) and the placenta.

The two hormones, oestrogen and progesterone, are carried in the mother's circulation and affect many different tissues in her body. They produce a number of important changes that help to provide the best environment for the growing foetus.

Because all the cells in the body are in contact with the blood, they will all be bathed by the hormones. Cells in the breasts are programmed to respond to increased levels of oestrogen, which increases breast size by developing the glands that will be needed to supply milk. In the womb, progesterone will relax the smooth

The sequence of photographs across the top of this spread charts four months of development. At five weeks, the beginnings of the spinal cord and the four limbs are visible; at seven weeks the fingers have formed and the face has a dark eye spot; at ten weeks the feet have toes; at eleven weeks – it's a boy! – a tiny penis is distinguishable; at four months the foetus is very babylike, tethered to the placenta by the umbilical cord.

Left, at five months of age, the down-covered face expresses an other-worldly peace – it is very comfortable floating in one's personal pond of amniotic fluid, dotted with tiny bubbles. Beneath the fused eyelids the eyes already have a lens and a retina – in the picture above the lens is at the top and the light-sensitive retina is the dark red crescent beneath it.

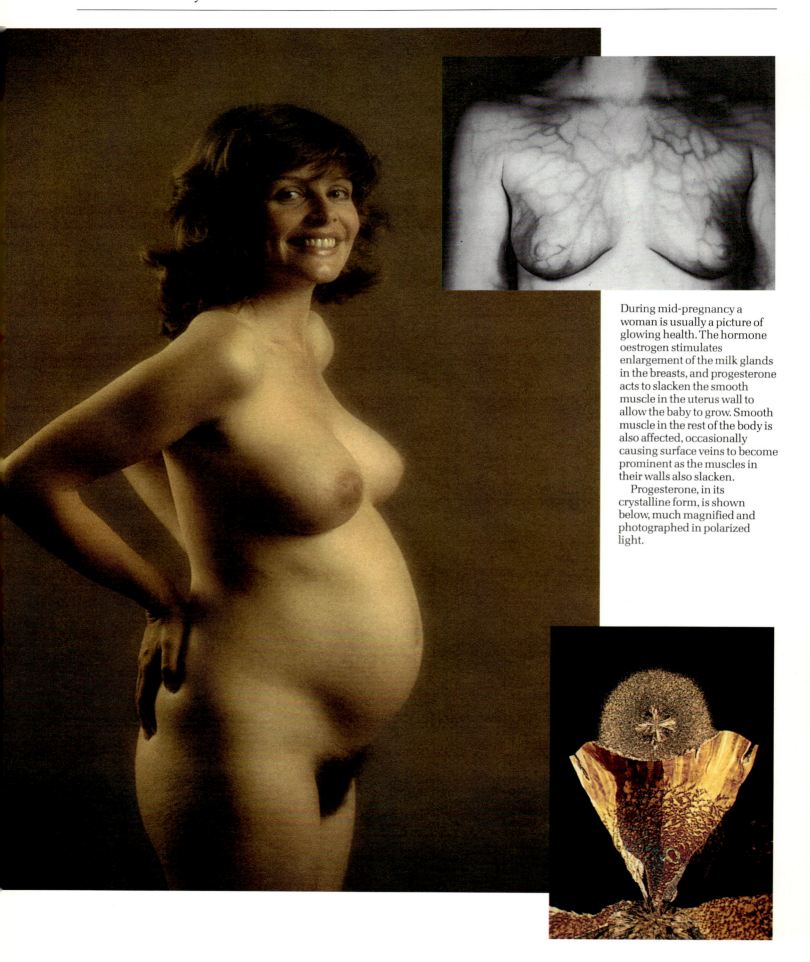

During mid-pregnancy a woman is usually a picture of glowing health. The hormone oestrogen stimulates enlargement of the milk glands in the breasts, and progesterone acts to slacken the smooth muscle in the uterus wall to allow the baby to grow. Smooth muscle in the rest of the body is also affected, occasionally causing surface veins to become prominent as the muscles in their walls also slacken.

Progesterone, in its crystalline form, is shown below, much magnified and photographed in polarized light.

Though it is not strictly necessary to 'eat for two', a pregnant woman does require more food. Usually she is advised to eat twice as much protein – 100g (3½oz) a day rather than 50g (1¾oz) – to help build the developing foetus. In general several small snacks a day are preferred to one or two large meals, partly because a full stomach becomes uncomfortable as pregnancy progresses. Morning sickness is best relieved by rest and small snacks of bland foods, such as biscuits, bread or bananas. Greasy or spicy food tends to aggravate queasiness or indigestion.

Weight gain during pregnancy is made up roughly as follows: the baby 40 per cent, placenta 10 per cent, amniotic fluid 10 per cent, uterus and breasts 20 per cent, blood and retained fluid 20 per cent. The usual pattern of weight gain is nothing or very little in the first three months, a quarter of the total in months four to five, half in months six and seven, a quarter in month eight, and nothing in the last month. Provided this is *broadly* the pattern there need be no cause for worry. A weekly weight check is a good idea though, to provide encouragement and evidence that all is well.

Staying active is one of the most important things a mother-to-be can do, both for her own health and the baby's. Maintaining family and social contacts is much better than spending lonely days at home. Walking is good exercise, and much better than standing. As bending and lifting become more awkward it is more comfortable to squat, kneel or get on all fours. Antenatal exercises are good too – they maintain tone in the muscles of the abdominal wall, back and pelvic floor.

muscle fibres to allow the womb to grow with the foetus.

Unfortunately, because they are carried all over the body, these hormones can sometimes also have the troublesome side-effects familiar to many pregnant women. Because progesterone acts on *any* smooth muscle, not just in the womb, it can relax the gut muscle which is similar. By relaxing the muscle just above the stomach, it can cause heartburn; and by slackening the muscles in the walls of the leg veins, it can lead to varicose veins.

The hormones also contribute to the growth of the foetus and of the placenta, the life support system that develops alongside it.

The placenta

The placenta is a much neglected part of childbearing, losing out to the far more glamorous and photogenic foetus. Dismissed as 'the baby's luggage' when it slips out after the birth, it is a versatile organ which fulfills a vital protective role for mother and baby. For the first weeks of life in the womb, the large flat mushroom shape of the placenta is far bigger than the embryo, which nestles insignificantly near the top of the womb. As the foetus grows, the placenta starts to take over some of the tasks that the mother's body has performed. By mid-term, when foetus and placenta share the entire womb between them, the placenta is performing a number of essential functions.

The stalk of its mushroom shape leads to the foetus and is the lifeline that carries vital materials from the mother to her baby. To reach the foetus, the mother's blood has to pass close to a fine membrane that has developed in the placenta, separating the two circulations. This acts as a filter through which some of the baby's waste products reach the mother's circulation, to be excreted by her. Although the placenta is only 20 centimetres (8in) across, it provides a much larger filtering surface because of the twists and turns of its intertwined blood vessels. If the filter were flattened out it would cover 10 square metres (100 square feet) – the size of a large carpet. Across it the foetus exchanges carbon dioxide for oxygen, and wastes for nutrients.

As well as filtering between the two circulations, the placenta produces its own hormones. It takes over the role of the mother's hormone-producing organs and concentrates on providing for the particular needs of foetal growth and development. The oestrogen secreted by the placenta may be the cause of the morning sickness that some women feel for the first three months or so of pregnancy.

Perhaps the most crucial role of the placenta is to keep the tissues of the mother and her growing child apart.

A month or less before it leaves the warmth and safety of the womb the baby adopts the typical head-down position ready for birth. The placenta, shown in the opposite page, is working to maximum capacity to provide oxygen and nutrients for the baby.

The baby is curled up and cramped, no longer able to somersault at will in the amniotic fluid. Its main movements are now leg kicks and elbow jabs, easily felt – and sometimes seen – by the mother. There may even be hiccups, from gulping amniotic fluid. All the baby's senses are now functioning. Awake, with eyes open, it sees light filtering through the thin, stretched tissues of the mother's uterus and abdominal wall. Loud noises, particularly low frequency ones, can startle it and make it jump.

The contour image on the right shows a pregnancy nearing its term. The weight of the baby places considerable strain on the mother's spine.

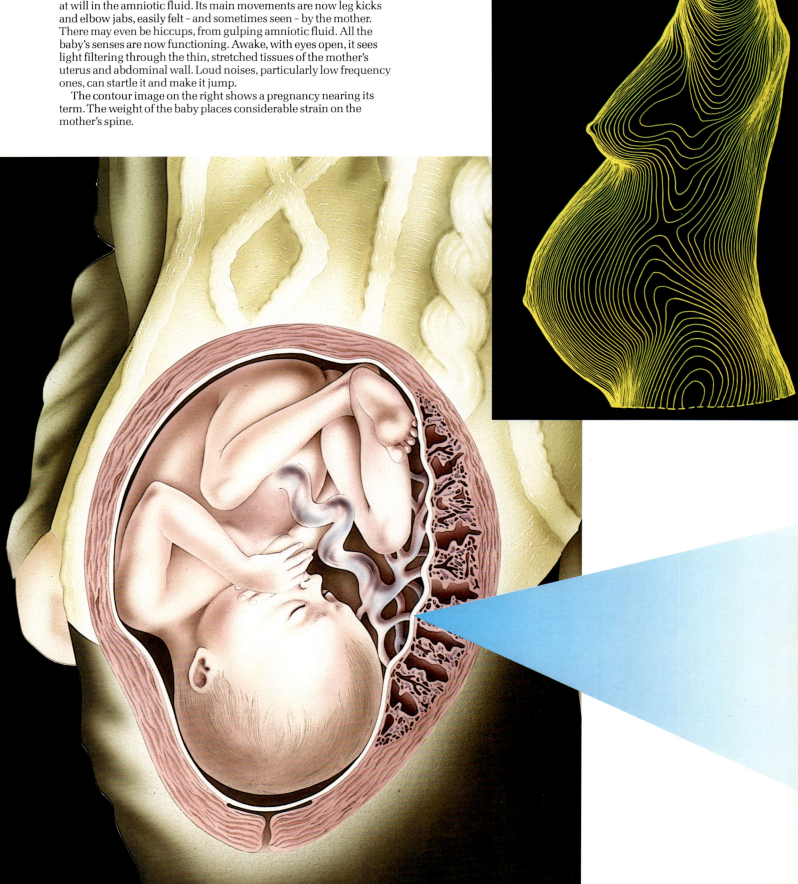

The foetus is foreign tissue in the same way as an organ transplant from someone else would be. Like a transplant, it could come into conflict with the mother's immune system which has evolved to prevent such material getting into the body. Although it shares the mother's body, the foetus is a different person, with a distinct genetic make up and body cells, half of which come from the father. If the foetus is male, this difference is very marked, with male genes and hormones which *should* produce rejection. That they do not is largely due to the effectiveness of the placenta in keeping the two tissues apart.

Later stages of pregnancy

As the pregnancy continues, the baby in the womb is able to do more and more for itself. By mid-term its organs have developed to the stage where they start to act on their own, making use of the nutrients and gases that come in the maternal blood. It grows at its fastest rate up to five months: if it continued to put on weight at this rate it would weigh 88 kilograms (190lb) at its first birthday! Its nervous system is developing so that it can begin to perform simple reflex actions, opening its mouth and withdrawing its limbs if they are touched. It also starts to use its muscles voluntarily to move around in what space is available.

At first the foetus has plenty of room to move about in the clear straw-coloured fluid, known as amniotic fluid, that helps to cushion it from the bumps and jolts of the mother's movements. This is not a static fluid – in fact, it changes every three hours or so through the foetus's remarkable ability to absorb the fluid through its skin into its circulation and so onward through the placenta into the mother's blood supply for excretion by

her. After five months or so, the baby tries out its swallowing mechanism by drinking the fluid at the rate of about 0.4 litre ($\frac{2}{3}$ pint) a day, and absorbing it through its gut into its circulation. It also starts to use its kidneys, and towards the end of the pregnancy the baby adds to the amniotic fluid by urinating in it.

While these later months of pregnancy may seem long and drawn-out for the mother, for the foetus – a feeling, moving, learning being – the watery world of the womb is far from boring. While there is little or nothing to see, this is not a silent world. The most prominent noise is, of course, the mother's heartbeat, and a few other bodily noises that occur nearby. The womb is, after all, surrounded by the mother's digestive organs which have been crushed to the top of her abdomen to make way for the womb. Also, loud noises from further away are probably audible, and the muffled strains of music could probably penetrate the cultural world of the womb.

During the last few weeks of pregnancy, the baby acquires new skills, such as the ability to perform chest movements that will be useful for breathing. Some of the baby's movements will be felt by the mother. Violent actions by the mother may be met by kicks from inside. A more gentle regular movement from the womb will probably mean that the baby has been at the amniotic fluid again and has hiccups.

Most women who have been carrying a baby for eight or nine months are more than ready for the birth. They have been supporting an extra 12 kilograms (26lb) for most of their waking hours – the baby will contribute less than one third of that, while the rest is made up of placenta and fluid, increased breast mass, and the extra fat the mother has acquired as a store for the job of supplying milk.

About 40 weeks after fertilization, the actual procedures of birth begin. From now on, mother and baby share an experience of some ferocity. As unstoppable as a roller-coaster, the events of birth can be both frightening and exhilarating.

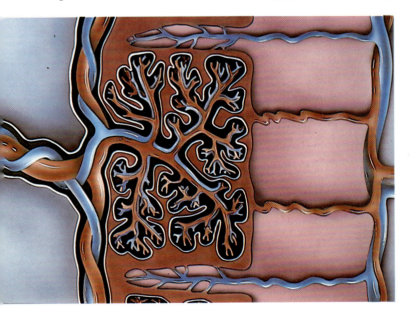

The cut-away placenta on the left is the baby's life support system. The two umbilical arteries carry deoxygenated, waste-laden blood (blue) into the placenta, they then branch repeatedly until they become fine networks of vessels floating in pools of maternal blood. It is here that the vital exchange of dissolved gases and nutrients occurs. Towards the end of pregnancy water filters in both directions between mother and foetus at the rate of over 3 litres; ($5\frac{1}{4}$ pints) per hour.

24 · Into the World

In the last chapter, we looked at how a baby in the womb develops from a tiny cluster of cells to a human being ready for birth. Towards the end of the nine months in the womb, the rate of development slows down. But there comes a day when, for no apparent reason, the tumultuous events of birth begin.

On that day, and for the next few weeks, the baby is introduced to a world full of new sights, sounds and smells. And from a state of total dependence, it has to learn the abilities it will eventually need to survive on its own in the world.

To prepare for the birth the mother's body has undergone many changes. By the time of the birth, the womb (or uterus) has expanded many times since conception. In the wall of the womb is smooth muscle tissue, making it the biggest muscle in the human body during late pregnancy. It has the job of pushing evenly and strongly in one direction – downwards – when it is time for the baby to be born. By then the vagina will have shortened to become the baby's gateway to the world.

Preparation for birth

In the last few days before the birth, mother and baby have begun to make final preparations. The baby has been making breathing movements for some weeks, although its lungs are surrounded by fluid and are virtually collapsed, and there is no air to breathe. While in the womb the baby obtains its oxygen through the placenta: oxygenated blood travels in the umbilical vein from the placenta down to the baby's body. But these breathing rehearsals are important, because at the moment of birth the baby will be cut off from its mother's oxygen supply and will have to use the diaphragm chest muscles it has been exercising.

There have even been rehearsals by the womb itself. During pregnancy, the smooth muscle of the uterus occasionally contracts on its own. Near the time of birth, these contractions become stronger and more frequent and sometimes act as a false alarm to the mother. They may reflect fluctuations in the balance of hormones controlling uterine muscle activity.

A few weeks before birth, the baby's head usually slips down into the pelvis, and the mother feels a lightening of pressure. It can happen that the baby's bottom slips down instead of its head. When this occurs, the baby may have to be born bottom first – a breech birth – although it may turn around a few days before the birth and emerge the most convenient way.

The pelvis is made up of three bones, which are normally held firmly together by ligaments. But pregnant women produce a hormone called relaxin which softens the ligaments and loosens the connections so that the bones can ease apart when the baby starts to leave the womb.

Nobody is really sure whose body triggers the birth – the mother's or the foetus' – although experiments in animals suggest that it is the foetus that sets in motion the events of birth.

The foetal pituitary gland releases a hormone, ACTH, or adrenocorticotrophic hormone, which seems to stimulate the womb through the release of another hormone in the womb itself. Once contractions start they are encouraged and strengthened by the presence of yet another hormone, oxytocin, in the mother's blood.

One of the first effects, and what is often the first outward sign that birth is under way, is the breaking of the waters. Up till now, the baby has been cushioned from the jolts of the world by the amniotic fluid that fills the womb. But with the first strong contractions, the weakest part of the womb – the exit towards the vagina – starts to open. The liquid is pushed against the wall of the bag that holds the baby by a strong contraction, the membrane gives way and the waters pour out.

The birth

Now birth really is under way, because the baby will not be happy or comfortable in a waterless womb. Its way to the world is now open and its journey must begin. Sometimes, as the muscular mouth of the womb stretches in preparation for allowing the baby out, some of the small capillary blood vessels break and there is blood mixed with the waters.

At first, these contractions come about every half hour, tightening up the womb around the baby, and gradually nudging it in the right direction.

For the baby, life is becoming less pleasant. It has never been more cramped. As well as being uncomfortable, the contractions limit the baby's oxygen supplies. Every time the muscle contracts, the blood vessels bringing oxygen through the wall of the uterus are squeezed shut. For a moment or two, the baby is short of oxygen. This does not matter very much when

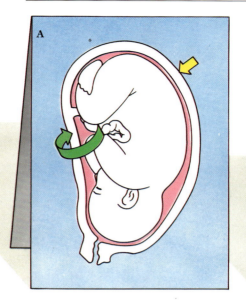

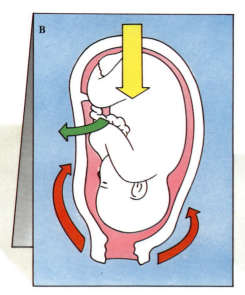

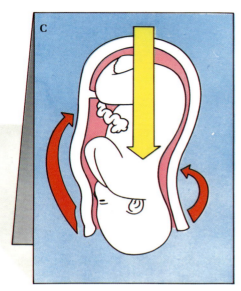

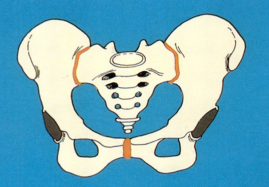

Labour is regulated by hormones. Towards the end of pregnancy (**A** above) the amount of progesterone (green arrows) in the womb decreases, allowing the womb muscles to be more responsive to continuing supplies of oxytocin (yellow arrows), which causes contractions. During labour, dilation of the cervix (**B** above) results in more oxytocin production and less progesterone production, so contractions increase. The presence of another hormone, relaxin (red arrows), then causes the mother's pelvic ligaments to loosen (left), allowing the baby to emerge (**C** above). More oxytocin causes any further contractions needed and helps the placenta to separate from the wall of the womb.

The relationship between a mother and her newborn baby develops from the moment of birth. After the first birth cry most babies settle into a quiet, semi-alert state punctuated by brief periods of fussing and crying. On the first day of their lives most babies spend about seven hours awake, but for the next week or so the amount of time they spend awake drops sharply.

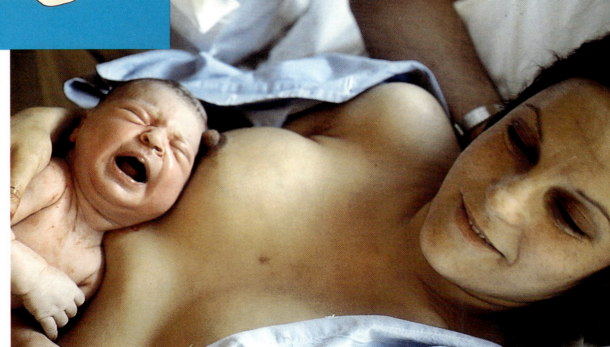

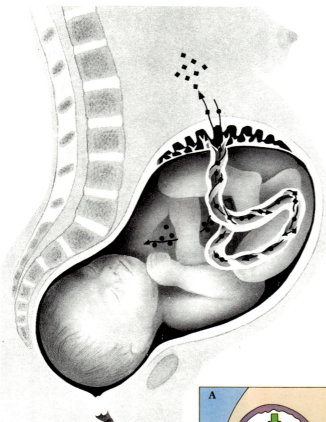

On the right is a coloured X-ray of a foetus at eight months, showing the head pointing downwards and 'engaged' in the mother's pelvis. Note how the skull nearly fills the pelvis. The baby's spine and legs are curled in the upper part of the womb, with the mother's spine behind them.

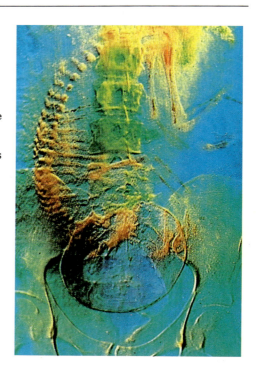

The placenta effectively keeps the mother's and baby's cells apart. Nevertheless all sorts of substances pass from the mother to the baby, including antibodies. If the mother's blood is Rhesus negative and her baby's is Rhesus positive, there is a small risk of the mother's antibodies attacking the baby's red blood cells, and causing some degree of anaemia. This risk is increased during labour, or following abortion and sometimes amniocentesis, when there can be some seepage of the baby's blood cells into the mother's. Injecting an Rh— mother with Rh+ antibodies ensures the few baby's cells that get into her blood are destroyed before she herself has time to manufacture antibodies. If she does form antibodies it is her *next* baby, if it too is Rh+, who will be at risk.

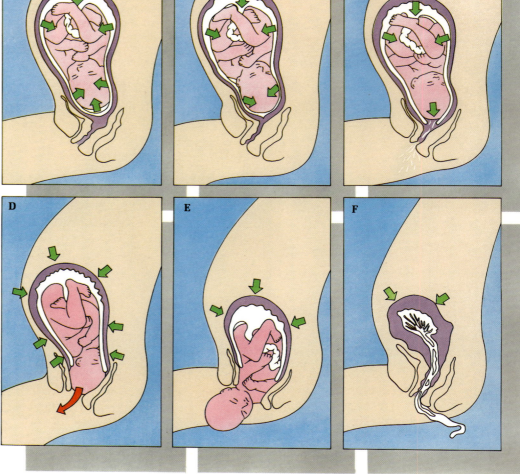

the contractions are once every half-hour, but the contractions increase in frequency as the birth draws nearer so that every few minutes, the baby's oxygen level drops and its carbon dioxide level rises. This contributes to the general stresses and strains, and the baby sometimes shows its distress by increases in heart rate.

As the uterus pushes from above, the neck of the womb, the cervix, opens out. In a non-pregnant woman, the cervix is a dome bulging into the top of the vagina. But at this point in the birth, the dome has disappeared, thinned out to a ring of muscle around a hole that gets larger as the baby's head moves down.

The baby, which was lying with its head pointing towards the mother's side, has now slipped round so that it is face down. It begins to slide through the gap in the pelvis, which has been specially relaxed to give the head as much room as possible.

Even so, it is a tight squeeze. Fortunately, the baby has a soft skull so that its head can actually squeeze into quite a distorted shape, without any damage. Also, it is made up of flat plates of bone that can actually overlap if necessary, if the passage is particularly narrow. After birth, during the first year, these bones grow to fill the two large gaps (fontanelles) between the plates, to become the protective covering for the brain. It will be some years before the bones harden properly. The top of the head, complete with hair, will be visible from outside by the time the cervix is seven or eight centimetres dilated. Now nothing can stop the baby's entry into the world.

For nine months the child has lain in the womb, anonymous. Boy or girl? Redhead or blonde? Blue eyes or brown? Suddenly the answers to all these questions, which were determined at the moment the egg was fertilized, become apparent as the baby emerges into the light.

The baby's birth is heralded by its first cry. It is actually the first breath that is important, but a lusty yell proves that the lungs are fully open and working.

Although the baby is often given a slap to help it,

At the onset of labour **A**, contractions of the womb act with equal force in all directions. With the first few contractions **B**, the cervix flattens and the baby moves downwards. The force of the contractions then begins to act downwards **C**, gradually causing the cervix to dilate, and the membranes that formed the 'bag' containing the baby to rupture. As the contractions become stronger and more frequent **D**, the cervix continues to open. When it is fully dilated – to about 10cm (4in) in diameter – the second stage of labour, the actual emergence of the baby **D**, begins. At this stage the mother feels a strong urge to push downwards with each contraction. Once the baby's head is born **E**, a few more gentle contractions are needed to ease out the shoulders and the rest of the body. In the third and final stage of labour **F**, the placenta separates from the wall of the womb and with a final contraction is expelled.

there are other events at birth that probably trigger the first breath. For the first time, the baby uncoils its spine and stretches out; its chest and diaphragm straighten up; and even the drop in temperature between womb and room can lead to a sharp intake of breath and a cry.

Behind this simple event lies a great deal of activity for the baby's blood circulation. During the last half-hour of labour, the baby may have become a little short of oxygen. In the first 30 seconds to one minute after birth, the baby has to switch from an oxygen supply that has been piped down its umbilical cord from its mother to doing something it has never done before – expanding its chest and sucking air into its lungs. At birth, the baby's oxygen supply is cut off as the maternal placental vessels begin to constrict. It can no longer rely on its mother's lungs to gather oxygen – it must breathe for itself. The baby's circulation has enough residual oxygen to last for a few seconds. After that, the baby is on its own. The lungs, which contain a special fluid that is removed at birth, must receive a full supply of blood so that oxygen can be taken on board and carried around the body. It takes several breaths to establish a normal breathing pattern, at which point the cord is clamped and cut.

Up to now, the flow of blood through the heart has been from the placenta, around the body, and back to the placenta. The two halves of the heart have acted together as one pump. Now, they have to act as two: instead of just pumping blood round the body, the heart has to pump it through the lungs as well. Most of the heart's internal plumbing will have to change its flow.

With the first breath, several important changes take place in the heart and blood vessels. Two major passages, the pulmonary arteries, begin to carry blood from the heart to the lungs so that the baby's blood can be diverted to pick up oxygen. There are also passages that have to be closed down because they are no longer useful. For example, there is a connection between the pulmonary artery and the aorta that helps to bypass the lungs during foetal life. When this hole fails to close properly, the condition is serious as the heart's two pumps are not entirely separate, and oxygen supply to the body is poor.

The final stage of labour is the delivery of the once-vital placenta. With one more contraction this slips out after the baby; the midwife (or doctor) is always careful to make sure that none of it has been left behind inside the mother, and she may give the mother an injection to make the contraction more effective. Any placenta remaining inside might prevent the maternal blood vessels from closing properly, resulting in serious bleeding.

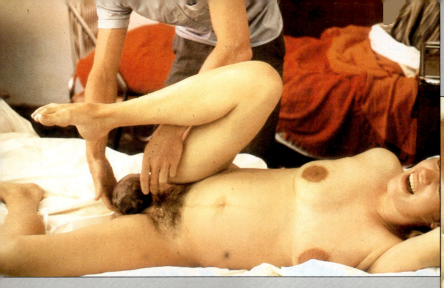

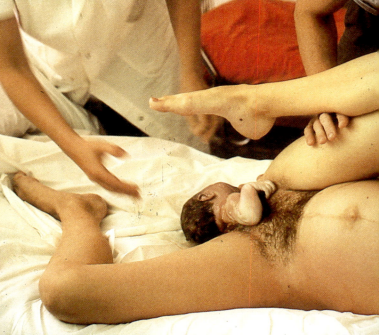

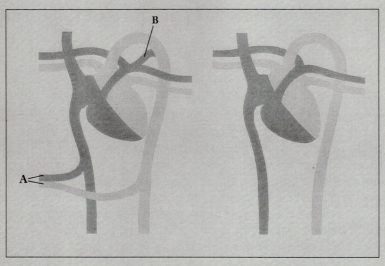

The two diagrams above show the changes that occur in a baby's circulation during and just after birth. The diagram on the left is 'before' and the diagram on the right 'after', and for the sake of clarity the vessels associated with the right side of the heart are shown dark grey and those associated with the left side light grey.

In the womb the baby receives blood from and returns blood to the placenta **A**; at birth the umbilical vessels are cut and their connections to the vena cava and aorta resorb. There is also a hole between the two atria, the foramen ovale, which closes at birth, and a duct **B** between the pulmonary artery and the aorta which enables blood to bypass the lungs, which are not yet functioning as blood oxygenators. This too closes, usually in the second week after birth.

By the time the baby's head emerges (above left), labour is almost over. Initially the baby is facing the mother's back, but in the next few moments he rotates to face her side. Then the arms, the body, and finally the legs gently slip out.

In most hospitals today, and certainly in all special birth clinics and when birth takes place at home with a midwife in attendance, the newborn is given to the mother to hold for a while. Then the umbilical cord is cut, and the baby is wiped clean and placed in a cot. 'Hello world, I want to get off . . . already.'

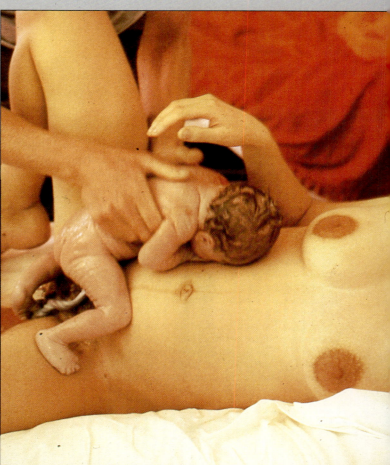

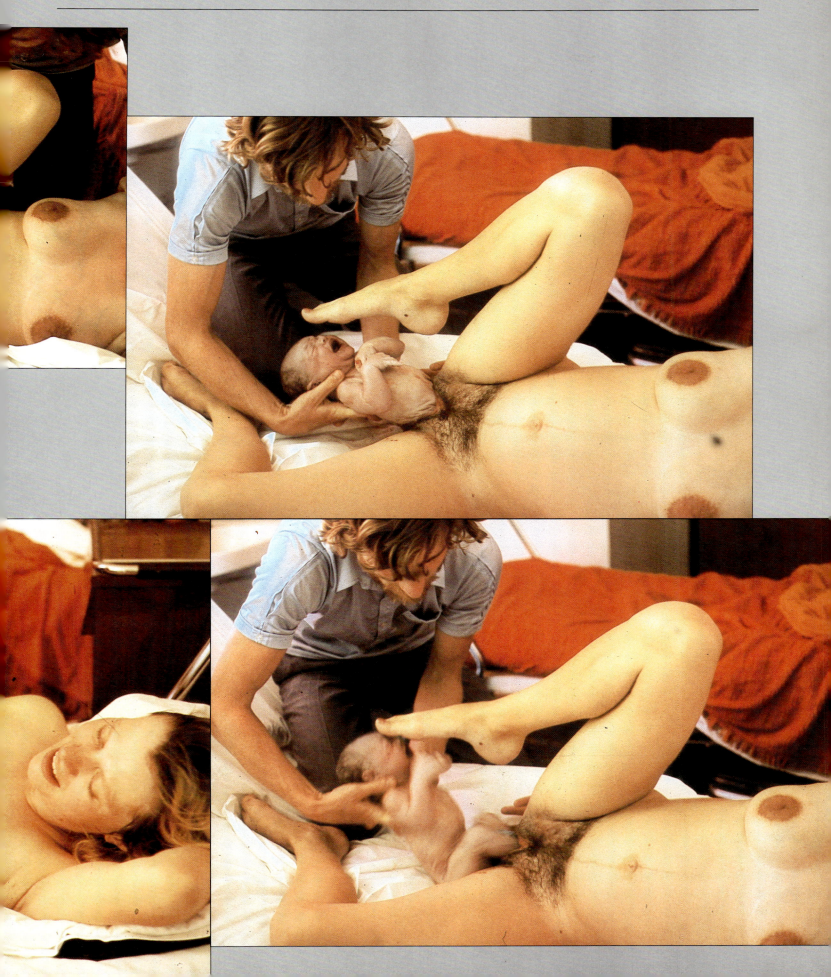

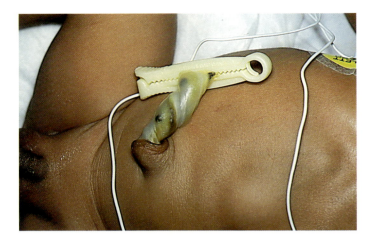

The baby's lungs begin to function with its first cry, but blood continues to pulse through the umbilical cord for another few minutes. When the blood flow has stopped, the cord is clamped to prevent infection, and then cut.

On the right is the expelled placenta. About 25cm (10in) in diameter and 3cm (1¼in) thick, the placenta detaches itself from the wall of the womb during labour and is expelled 10 to 20 minutes after birth.

We are born equipped with many reflex abilities. Almost immediately after birth, we can make walking movements (left) provided our weight is supported, move our arms and legs suddenly in coordination with each other (below), and instinctively grasp an object if it is placed in our hand (right). We can also direct our gaze towards sounds, and a little later turn our head towards them too; we can also turn our eyes and head to track a moving light source.

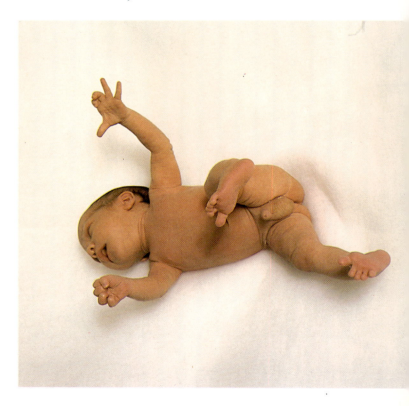

Events following the birth

Once the baby has taken its first breath and had a look around it is time to try out another system it is going to depend on for the rest of its life – the digestive tract. Up to now, the baby's meals have all been obtained by a direct infusion into its blood. Now it has to use its mouth, stomach and gut to get food from its mother.

This is one of the remarkable reflexes that a baby is born with – a skill it does not need to learn. By touch, smell and sight, it recognizes the importance of the nipple. Any light touch to a newborn baby's cheek will immediately cause it to turn its head towards the object. And when its lips find it the mouth automatically takes much of the nipple into it, so that the milk outlets are near the back of the throat. The automatic squeezing movements that the baby makes then trigger activity in the mother's breast to produce milk – another reflex action, this time in the mother's body (see page 130).

There is an important by-product of breastfeeding that is often overlooked. The need to eat in this way at regular intervals makes sure that mother and baby spend a lot of time together, familiarizing themselves with the sight and touch and smell of each other. In fact many psychologists look on the mother-baby bond established at this time as training for healthy emotional and physical relationships in adulthood.

Within its nervous system there are already nerve circuits which produce patterns of behaviour that the baby is going to need. The ability to clutch an object like a finger and cling onto it is already there. The startle reflex, when all the limbs are flung outwards at once, requires a lot of coordination and has never been learnt. Most remarkable of all is the walking reflex: within minutes of its birth, a baby can make walking movements, lifting one leg and then the other, if its weight is supported.

From now onwards, these simple reflexes will be joined by other types of behaviour. The built-in nerve circuits will be added to and modified. Whole new sets of connections will be made in the brain. And this learning goes on at a furious pace. At birth, the baby's brain is already one quarter of the weight of an adult brain, whereas the rest of its body is only five per cent or so of its adult weight. During the first six months, with all the learning it will do, its brain will double in weight, to reach 50 per cent of its adult weight.

The act of birth is neither a beginning nor an end for the baby. The development that began when the egg was fertilized continues for many years after birth. When the baby arrives in the world it is merely making a transition to a more challenging and stimulating environment in which to grow to maturity.

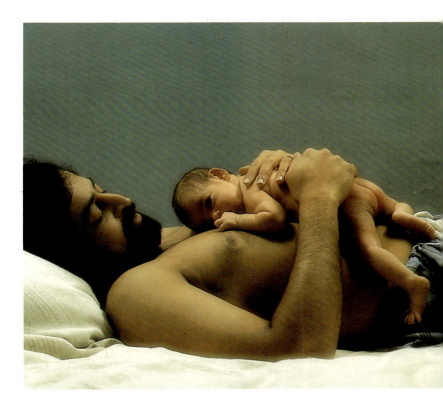

A great deal has been written about the mother-baby bond, but very little about the bond between father and baby. Nevertheless, most child psychologists agree that babies thrive on affectionate contact, skin to skin, whether or not the object of the exercise is feeding. The support that a father can provide is more than emotional.

25 · Life Before Death

If you look carefully enough, you can find the signs of age in every part of the body as it nears the end of its life. Every organ and system seems to be bound up in a process which starts subtly at a surprisingly early age but only really makes itself felt when we are too old to reproduce. The image of old age is universal. The Japanese word-symbol for old has not changed much since it was derived from the image of an old man: slightly bent, with a walking stick to give support, it depicts an old man with the same physical signs of ageing that will appear in all of us sooner or later if we are lucky enough to live to a ripe old age.

But why do we age and die? After all, it seems to conflict with the drive to survive that underlies most bodily activities. Body systems have evolved to collaborate in all sorts of intricate ways to prevent injury and death from the hostile elements in the world around us, so that we can perpetuate ourselves in our children. We use our immune systems to fight microscopic attackers and our intelligence and nervous systems to plan the safest path around and away from the threats of animals and our fellow humans. And yet, as we shall see, there appears to be a plan for death as well as survival. It seems that we *have* to die for evolution to continue, and before we die we *have* to age.

There is a wide variation in the effects of ageing on different individuals. Some women of 60 appear 10 or 15 years younger, some men of 80 appear no older than 50; and yet there are others who seem old before their time – mentally and physically worn out at a time when they could still have years of active life left. Some of the differences are due less to ageing than to what might be called wear and tear. States of mind can have a part to play as well. But whatever the state of mind, there are certain things which will happen sooner or later in every human body.

Physical effects of ageing

Some of these physical signs of ageing are familiar because they are so visible. Grey hair, for example, is one

The physical effects of ageing vary greatly from individual to individual. Some people are still capable of strenuous gardening or even more vigorous exercise at the age of 80 – in full control of their limbs, and able to keep their balance and lift heavy weights as easily as they did in their youth. Others may suffer from deteriorating eyesight, painful joints and loss of coordination at 60 or earlier.

There is also wide variation in the rate at which people age mentally. Although physical condition certainly does affect self-esteem and self-image, many elderly people become the victims of society's generally dim view of their capacity for enjoyment, creativity and learning.

Among the more conspicuous signs of loss of youth are grey hair and baldness. Hair turns grey and then white because cells at the base of the hair, in the hair follicles, stop producing the pigment melanin. Gens and hormones determine at what age and with what speed a person goes grey, although illness, shock or extreme stress can accelerate the process.

Old age does wither us, inevitably. The collagen fibres that support the skin – the photograph below shows what they look like under the microscope – become stretchy and skin loses its firmness. Loss of fatty tissue under the skin also accelerates wrinkling. So does regular exposure to hot sun, though a tan always gives an impression of youth and vigour. The youthful 65-year-old on the right is also likely to be 5cm (2in) shorter than she was in her 20s; this is because our posture-supporting muscles and the discs between our vertebrae shrink as we get older.

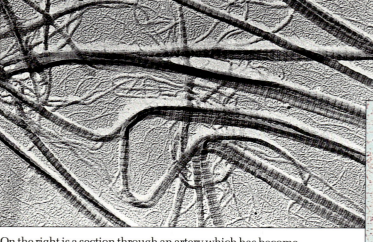

On the right is a section through an artery which has become dangerously narrowed by a large deposit of atheromatous plaque – a mushy mixture of scar tissue, cholesterol and calcium salts. One of two unpleasant things might happen, either in the artery itself or elsewhere in the circulation: blockage or bursting.

If the deposit continues to build up – and as it builds up it destroys the muscles of the artery wall underneath – the artery may bulge and burst with the pressure of blood in it – an aneurysm. If the deposit gets dislodged it may travel elsewhere and block a narrower blood vessel; if it blocks a blood vessel in the brain the result will be a stroke, and if it blocks a coronary artery the result will be angina or a heart attack. Or a clot may form at the site of the detachment and either block the artery at that point of travel round the body like a time bomb – sooner or later it will cause a fatal blockage.

of the earliest and most visible signs of age. The hair generally keeps growing until we die, but one group of cells in the skin, responsible for hair colour, can stop working much sooner, usually in our forties or fifties, or even earlier in a few people. These cells no longer produce the brown pigment, melanin. This is the pigment responsible for producing a suntan and it also occurs naturally in the hair in varying amounts, depending on whether the hair is dark or fair. At a certain age – earlier than some people would like – the melanin-producing cells stop working. Although the cells that produce the hard outer layer of the hair continue to push out new layers from the skin, these new cells are now transparent, giving a grey appearance to the hair.

Beneath the head of hair is an ageing brain. Here, the sad news is that silently and invisibly it has been getting older from the moment it reached maturity. Every day, each of us over the age of about 20 loses 10 000 or so brain cells, and during our natural lives we will lose 180 million or so by this stealthy and unobserved wastage. To put this figure into perspective, we are still left with about 10 000 million to work with during old age – so we only lose about one fortieth of our total.

Nevertheless, the loss of brain cells can have significant effects on our mental abilities, and it may well be that some parts of the brain lose a relatively greater proportion than others. If we analyse the brain shrinkage in older people, comparing the brain size with the skull cavity, it seems that between the ages of 20 and 60 we lose about one per cent of our brains, while

between 60 and 99 we could lose another 10 or 11 per cent. (There is no direct correlation between decrease in brain size and loss of brain cells.)

These losses can make themselves embarrassingly apparent in the memories of older people, although other factors may be involved. Familiar names can no longer be recalled easily and recent events are not always remembered as clearly as those of youth and childhood. Part of the reason for this may be that for older people recent experiences are inevitably less novel than earlier ones – 'they have seen it all before' – and so are not likely to have the impact of youthful events.

Another well known sign of ageing is wrinkling of the skin. Just under the surface of the skin lie elastic fibres of connective tissue called collagen, which provides support and toughness. As we get older, the collagen actually becomes more elastic, with the result that it provides less support for the skin, allowing it to slacken and pucker into ridges and folds. One important cause of this change is exposure to ultraviolet light, and wrinkling therefore occurs most and earliest in those parts of our bodies that are exposed to the sun. This can be clearly seen in the differences between two people of the same age, one of whom has worked most of his life in the sun, and the other who has led a much more sheltered life.

The blood vessels are also commonly affected by ageing. If the arteries of lots of old people in a population are investigated many of them may be found to have hardened and narrowed – atherosclerosis. With this condition, the walls of the blood vessels have lost a lot of their elasticity and have become stiffer. In addition, there are deposits on the walls which narrow the passage and make it more difficult for the blood to pass through the blood vessel. This narrowing of the arteries is partly related to the type of food we eat: an excess of fat in the diet can lead to a 70 to 80 per cent reduction in the diameter of some blood vessels.

But although atherosclerosis occurs in many old people, is it a natural consequence of old age, like grey hair or wrinkles? In fact it is not. Most other mammals age without getting atherosclerosis; and throughout the world there are many populations whose people live to a ripe old age without having their arteries affected in this way. This is an example of an age-related disease – a condition that is not inevitable and is, indeed, preventable. It just happens that the longer we live the more chance there is for a bad diet to harden and narrow our arteries.

True effects of ageing are to be found in the muscles and bones. Older people have less muscle tissue than when they were younger, and since we are born with all

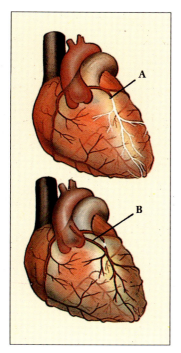

A partial blockage in one of the arteries that serve the heart causes angina, a pain in the middle of the chest, and sometimes in the neck or under the jaw and down the left arm.

The drawings on the left show the surgical solution to the problem – a coronary bypass. A small section of vein is taken from the robust saphenous vein in the leg and grafted between the aorta and the part of the artery beyond the blockage. The vein has to be turned upside down so that its valves will not block blood flow. **A** marks the blockage that is starving part of the heart of oxygen, and **B** marks the bypass.

Non-surgical treatment would involve stopping smoking, avoiding rich or fatty foods, losing weight, taking regular and moderate exercise, and avoiding stress.

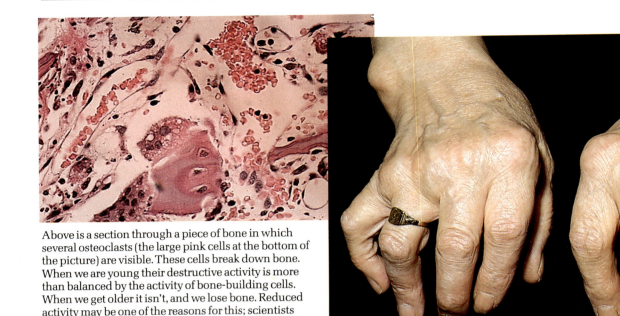

Above is a section through a piece of bone in which several osteoclasts (the large pink cells at the bottom of the picture) are visible. These cells break down bone. When we are young their destructive activity is more than balanced by the activity of bone-building cells. When we get older it isn't, and we lose bone. Reduced activity may be one of the reasons for this; scientists returning from Spacelab have been found to have lost bone during their periods of weightlessness.

On the right are the swollen knuckles typical of rheumatoid arthritis, inflammation of the fibrous connective tissue around joints.

the muscle fibres we are ever going to have, there is nothing we can do to replace the ones we lose.

There is, however, a ray of hope. In spite of the inexorable downward path of our muscular abilities, we can still slow down the rate of deterioration by exercise and training. Anyone of any age can improve his physical condition by using his muscles, and the muscles that are still available can be strengthened by regular and moderate practice.

The heart is a special sort of muscle. It has its own built-in control system that helps the body to respond to the changing demands of physical activity. As the heart ages it seems to lose some of its versatility: it takes longer to react to changes in demand; it pumps more feebly and with less coordination; and it has fewer reserves of power to call up suddenly when needed. This means that when life is not strenuous the ageing heart is perfectly capable of doing its job, but when faced with the need for a sudden increase in effort, in a stressful situation that would not worry a younger person, it will not be able to cope so well.

Such stresses can contribute to the causes of heart failure, a condition in which (despite its name) the heart does not stop completely – as happens in a heart attack – but pumps much more weakly than a normal heart. This condition is likely to be aggravated by atherosclerosis of the arteries supplying the heart with blood.

Like the muscles, the bones deteriorate with age. As well as providing rigid support for the body, the bones are an important store for minerals, particularly calcium, that the body needs to draw on from time to time. Here the changes as we get older are very clear-cut: we actually lose bone mass and this affects women more than men. The average woman loses between a quarter and a half of the substance of her bones as she gets older. Although they maintain the same shape, they become lighter and therefore weaker.

As we saw in Chapter 7, during the first half of our lives there is a constant balance between the skeleton's twin jobs of supporting the body and acting as a mineral store. One group of cells liberates calcium from the bones while another group concentrates calcium to turn it back into bone. In the midst of this turmoil there is always enough free calcium to be used by the other parts of the body. Part of this process is controlled by hormones, and after the menopause, when oestrogen levels drop, women's bones lose some of their calcium. But part of it is affected by the level of stress imposed on the bones by physical activity. The parts of the bones subject to the greatest pressure remain the strongest, and bones which take less stress become weaker.

In addition, the balance of activity in the bones changes as we age, more calcium being liberated from the bones than is being replaced. There is therefore a loss of calcium from the bones into the rest of the body. Although all bone looks like a honeycomb, the older it is the bigger the holes are. This means that old bones are not as strong as they used to be and cannot take such

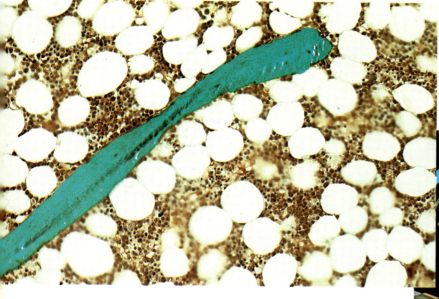

On the left is porotic bone, bone that has become weak and brittle through loss of calcium. The bony spicule, seen here as a blue strand, is very narrow and weakened and surrounded by soft marrow which has taken the place of other spicules of hard bone. A bone in this state would be vulnerable to the slightest knock.

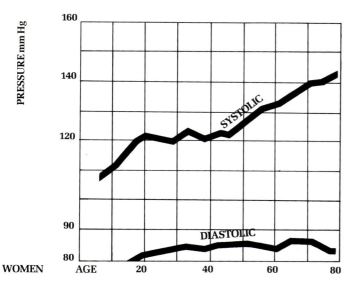

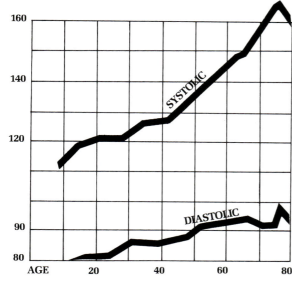

Osteoporosis, or decalcification of bones, particularly affects women after the menopause. Partly this is due to the fairly sudden drop in oestrogen production – crystals of oestrogen are shown at the top of the page – that marks the end of a woman's reproductive life. But partly it is a consequence of the habits of a lifetime. The old lady above, looking for worms on the seashore, has led an active outdoor life since girlhood; she is less likely to become disabled through brittle bones than a woman of comparable age who has never paid much attention to exercise.

As we get older our blood pressure increases, and with it the risk of heart and kidney ailments and strokes. In young people of both sexes systolic pressure of 120mm Hg and diastolic pressure of 80mm Hg are the norms. But as arteries age and become less elastic blood pressure rises. The average 60-year-old man has a systolic pressure of about 140, and by the time he is 80 this will have risen to about 160. At all ages, women tend to have lower blood pressure than men.

hard knocks. Unfortunately, that is exactly what they have to put up with – one of the consequences of the deterioration of nerves and muscles that accompanies ageing is that control of movement is less precise, so older people are likely to suffer more knocks and jolts. These, in turn, put more strain on weakened bones, increasing the probability of serious fractures.

Another effect of knocks and jolts is to put damaging pressure on the joints. If the bones themselves are not crushed by these pressures, then the joints have to take the strain and they become damaged instead. Many old people suffer from osteoarthritis – a disease in which the cartilage of the joints degenerates, causing pain. It seems that old people have to put up with one or the other, damaged joints or damaged bones, because of their more erratic and unsteady movements.

Another important organ that begins to fail as we get older is the lungs. The lungs of an 80-year-old have lost half or more of their ability to contain and pass on oxygen. This means that, while the blood of a 20-year-old man can take up 4 litres (7 pints) of oxygen per minute, an 80-year-old can only take up 1.5 litres ($2\frac{1}{2}$ pints).

Similarly, the kidneys lose about half their efficiency over the last 50 years of life. As with brain cells, the hairpin-shaped tubes that exist in their millions in the kidneys die off in large numbers as we get older.

Delaying deterioration

This tour of the ageing body is hardly encouraging for any of us. The catalogue of organs and systems crumbling quietly away may seem a dismal one. But, of course, to mention every possible sign of ageing in every possible organ is not giving a true picture. There is much individual variation – some people will have lively and alert minds until a ripe old age, even if they are not so active; others will still stride out at 80 even if their memories are not as good as they used to be. And still others will retain healthy, functioning minds and bodies until the day they die.

With some systems at least it is possible to halt or prevent the decline. It may be that bones become weaker as a result of the decreasing level of physical activity of older people. Because their hearts and muscles are less useful, they tend to become less active. This, in turn, puts less stress on the bones, which become weaker as a result. It is at this point that sudden increased effort can have disastrous consequences for a long bone, like the thigh, which has become more brittle. But if the level of physical activity has always been kept high enough to keep the pressure on the bones, then at least some of the bone loss will be prevented and fractures will be less likely to occur.

In a similar way, the heart muscle can be encouraged to retain some of its power by being used. A gentle and regular increase in activity will keep stretching the heart muscle, which will increase in size and strength, so it may then be able to cope with a sudden emergency demand for activity – the sort that would cause a weaker heart to fail.

In the end, of course, the body inevitably loses some of its abilities with the passage of time. All the evidence suggests that this is not just a by-product of human biology – a problem that nature has been unable to solve – but is actually part of its design, a sort of built-in obsolescence. Each cell type in the body seems to have a set limit to its abilities to reproduce. This natural lifespan of individual cells is maintained even outside the body. If you were to take some living cells from your own body, from the inside of your mouth, say, and culture them in a dish, they would start dividing vigorously to produce a growing colony of identical cells. But sooner or later, however well you fed and looked after them, they would stop dividing. The final generation of cells would lose the power to divide.

With any cells from the human body this loss of reproductive ability happens on average after about 50 divisions. They seem to be programmed to stop at a certain stage. Although this occurs in the artificial conditions of a laboratory, there is evidence to suggest that it is linked to the processes of ageing, which occur in the living body.

This limit of 50 generations is true for cells from a foetus or baby. But the cells from older people stop dividing sooner, suggesting that they have a built-in clock that says it is time to stop now – you have had your allotted three score years and ten. Similarly, if you transplant some body cells from a young animal into an older one they can help the older one live longer, because their internal 'clocks' are set to an earlier time. And it is also the case that the longer-lived a species is, the more divisions its cells will undergo if cultured in the laboratory.

However fast or slow the pace at which we decline, there comes a moment for all of us when the living body stops living. As we have seen, barring accidents, there is a gradual and parallel decline in the different organs and systems of the body. If all the major causes of death were eradicated, the human body would live for about 100 years, with a maximum lifespan of 130 years. The end would eventually come because, as functions fail, our body systems are working nearer and nearer to their minimum effective level. The heart, the lungs, the kidneys and the liver have fewer and fewer reserves to summon up when the occasion demands. The repair mechanisms become increasingly less efficient. The only protected organ is the brain. Although it also deteriorates, it will not fail abruptly if it suddenly has to do a lot of thinking. Right to the end, unless it is affected by disease, the brain, the essential centre of our humanity, retains its personality but, as always, it is dependent on the support and cooperation of the rest of the body. Natural death occurs in the brain but is not usually the brain's fault.

There comes a point at which the failing ability of the heart falls below the level of activity required of it. Then the heart stops beating. Suddenly, all over the body, the blood stops flowing, unable to unload any more oxygen or take on any more carbon dioxide. Each group of cells in its own colony is suddenly suffocated, but some of them can do without oxygen for longer than others.

Some cells start to die immediately – the brain cells, for example, have a maximum survival time of seconds. Other cells can last much longer without oxygen. The muscle of the gut, for example, preserves its integrity for a considerable time after all circulation has ceased. If cells are in the middle of dividing when death occurs they finish what they are doing rather than stop in mid-division. And in a man, the vasa deferentia, the tubes that carry the sperm from the testes, seem to be one of the last types of tissue to retain some sign of life.

If current ideas on ageing are correct, then death in old age occurs when one component of the body fails in a body where all parts are nearing the end of their useful lives, and it is a matter of chance which part fails first. The Chinese used to talk of 'death', when a young person died and of 'end of life' for the aged. Perhaps they would understand the futility of trying to fight natural death with whatever weapons medicine can devise.

Beyond the third decade of life most of us feel younger than our official age and this remains true of almost all of us right up to the end of our lives.

Though there is a reduction in the size of muscle fibres throughout the body, muscles remain in basically good condition into advanced age. Perception of body positioning and movement become less acute, but this is partly the result of sedentary habits. So not all the satisfactions of old age have to be philosophical; moderate physical activity, which feeds back into mood and general vitality, remains possible for most of us.

A lift for the face and a lift for the ego. But the effects of lifting are temporary, at the most seven years claim some clinics. And there is a limit to the number of times – up to three according to some clinics – that lifting can be done. The photograph above shows the 'before' and the 'after'.

26 · A Future for Man

We are used to thinking of ourselves as being at the peak of evolution – the final product of millions of years of living in a changing environment. We have seen in this book how the improvements of successive generations have led to the development of a body which is capable of impressive physical and mental achievements. But in fact we may not be at a peak: we could be continuing to climb. There is certainly no reason to think that we have achieved perfection – there is plenty of room for improvement. But are there eventual limits? How can evolution continue to operate in a world where humans are learning how to manipulate their own genetic material instead of leaving it to the much slower process of natural selection?

We are, of course, living in a different world from the one that shaped the human body. Nowadays, when we stretch our respiratory system or circulation to the limits it is unlikely to be because we are trying to save ourselves from a predator. Sharpness of vision and acuity of hearing are no longer essential for our survival. Even reproduction no longer needs to proceed at the hectic pace that used to be necessary to maintain or increase the population in the face of high infant mortality. In all of these respects, it is play rather than survival that is likely to stretch our bodies to their limits.

So to look at the future of the living body we shall have to make some assumptions about the conditions in which men and women will live and the pressures that will operate to change their bodies.

One way of discovering the future is to trace the changing events of the past, and see where they might lead if trends continued. When these changes produced a new organ it was always no worse than the previous version and was usually better in some way. The improvement usually meant that the creature could cope better in the world. If a designer wanting to make improvements had looked at the eye of a fish, he might well have said that it would do better if it had colour

In the last decade or so we have discovered ways of reversing decisions made for us by accidents of biology. For example, we can freeze sperm and ova, making it possible for infertile men and women to have offspring; we may soon be able to separate Y-carrying sperm from X-carrying sperm, and so be able to choose the sex of our children; we can freeze embryos, and so choose when to have children.

The photograph below shows batches of human sperm frozen in liquid nitrogen; on the left are human embryos being prepared for deep-freezing. Embryos are obtained by fertilizing mature eggs with sperm, freezing them until needed, and implanting them in the womb of the biological mother or a surrogate. They then develop as a normal pregnancy.

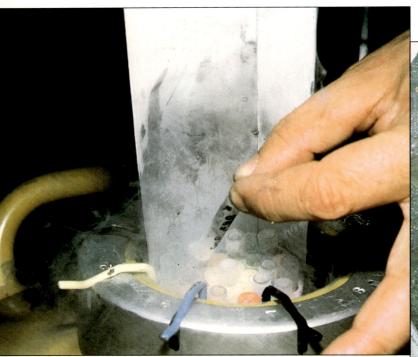

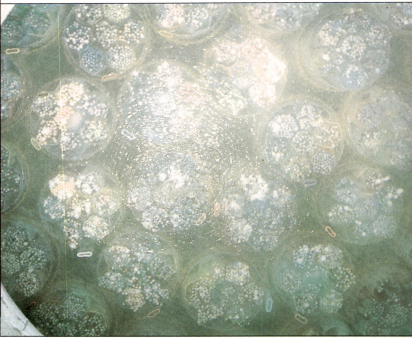

vision, a variable aperture and stereoscopic vision. Eyes eventually did acquire these refinements, and became much better seeing instruments.

But to do the same thing with the current model presents problems. By extrapolating from the changes from fish through reptiles to mammals, it might be possible to hazard a guess that in a few million years the eye will, perhaps, detect a wider range of radiation and have a zoom lens. But do such improvements have to be the result of chance genetic changes? No, not as far as we humans are concerned. We have bypassed this mechanism with spectacles and contact lenses, meaning that people with a far wider range of eyesight can survive and even triumph in the world.

This may mean that our individual organs will no longer evolve, because they do not need to any more. We live in a world where survival no longer depends on physical competition, and where people can live and reproduce successfully even if their organs and systems fall short of perfection. To go back to the example of the giraffe's long neck, in Chapter 1, if the short-necked ancestors of giraffes had invented stilts or ladders the long-necked variety might never have evolved.

Even if the physical structure of our eyes never changed in the future, our vision could still be improved by artificial means. Our descendants may be able to see infra-red, ultraviolet and X-rays, but if they do it is more likely to be with sophisticated add-on devices, using essentially the same eyes as we have today. It is our ingenuity that must continue to flourish.

But although we may not be very sure whether individual organs or systems will evolve in the future as they have done in the past, there are other aspects of the living body that could well change to our advantage.

Many people think that it would be a distinct improvement if we could live longer. The idea of immortality is, perhaps, too fanciful; it would also present a problem of overcrowding. Even a significant increase in longevity could have a dramatic impact on the population. A couple who married in 1800 and were still alive today could now have over 20 000 living descendants.

It is easy to think of reasons why too long a lifespan might not be an advantage to the human race. It could lead to a situation in which there was much less need to replace individual members of the population, and reproduction would play a much less important role; this in turn would lead to fewer opportunities for mixing genes, and less chance for natural selection to improve the species. And a world in which the population contained an increasing proportion of people whose bodies had many of the inevitable consequences of advanced old age might be a less than enjoyable place to live in.

Arrays of radio telescopes, below, have given us a faculty that unthinkable millions of years of random genetic shuffling would never achieve. But if the technologies that enable us to compensate for our mental and physical limitations go on getting better and better, will there any longer be selection pressures on us to evolve further?

Advances in medical science – the photograph below shows white cells being cultured for drug research – mean that we humans are, to an increasing extent, exempt from the law of survival of the fittest. Even those of us with less efficient internal defence systems are now surviving to pass on our genes.

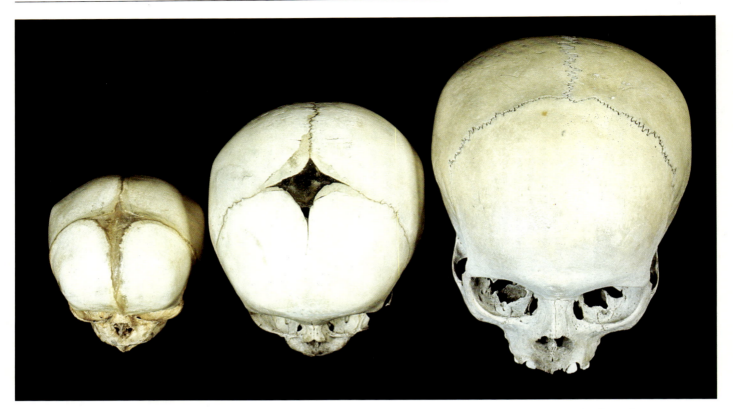

The skull plates of a newborn baby (left) are separate and flexible – birth would be very difficult if they were not. At eight months (centre) the skull has grown a great deal but there is still a gap between the plates. At five years old (right) the plates have come together, but the joins do not fuse and harden until adulthood.

An intriguing theory has been put forward, however, by the American anthropologist Ashley Montagu, that would involve the evolution of much longer-lived humans in the future, while avoiding the problems of large numbers of people who were 'aged' in the conventional sense. We could be progressing towards a world eventually populated by a very different type of human being, although one whose body was very similar to our own.

Instead of considering the evolution of individual organs, Montagu (and other scientists before him) looked at a less specific characteristic of humans – the way we develop, from fertilization to maturity.

If we analyse our developmental processes in comparison with those of our ancestors and of other animals, it becomes apparent that many animals are born with much more independence than the human baby. In relation to their total lifespan, they reach maturity much sooner, they lose 'childlike' physical attributes much more quickly, and they become able to leave their parents and start reproducing much sooner than we do. These observations led to the conclusions

that humans develop more slowly than their ancestors did and that adult humans preserve many features of the human foetus whereas adults of other animals soon lose their foetal characteristics. We preserve childlike facial features for the greater part of our lives; we do not have much bodily hair in comparison with our ape relatives such as chimpanzees; human flesh and skin stay 'babylike' for much longer; we mature sexually at a much later stage; our cranial sutures, the meeting points of the bony plates that make up our skull, do not close until adulthood, while in apes they close in early childhood.

It seems clear from all this that we retain into adulthood many features that are typical of the chimpanzee foetus or baby, but which the chimpanzee loses before growing up. A human newborn's skull is very similar to the chimpanzee's, as well as being similar in proportion to a human adult's. But there are great differences between an adult human's skull and that of an adult chimpanzee.

An example of the advantages of staying foetal longer is the size of the human brain. If we developed in the womb to the stage of maturity that many animal foetuses attain, our brain size would be such that the head would be too big to slip through the mother's pelvis. Because we leave the womb at an earlier stage, the brain can continue to develop for longer, and we gain all the advantages of increased brainpower. And

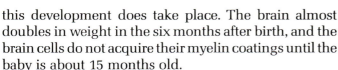

A chimpanzee's face changes a lot between babyhood and adulthood – the skin darkens, the eyes become proportionally smaller, the jaw protrudes, the brows beetle. A human face, by contrast, changes much less in the 20 or so years between birth and independent adulthood.

this development does take place. The brain almost doubles in weight in the six months after birth, and the brain cells do not acquire their myelin coatings until the baby is about 15 months old.

One result of this slow development has been the greater dependence of human young on someone else, usually the parents, until maturity. Most animals are much less dependent on their parents. Even animals with a gestation period as long as or longer than humans become independent soon after birth. Newborn elephants, rhinoceroses or camels, for instance, run with their parents shortly after they are born.

Such dependence might seem a disadvantage, but Montagu and others point out that it could be the care and protection which the young human receives from its parents that have led to much greater opportunities for learning, for creativity, for play and for flexibility. And all of these are traits that are characteristic of the way human beings respond to the world and make the most of their opportunities in it.

This close parental involvement in caring for the child is encouraged by some of the foetal characteristics that human children retain. The helpless newborn baby is an image that we find very appealing, and so the baby arouses protective feelings in the adults around it. It is suggested that the curved surfaces and soft skin of a baby are in themselves a cause of the bonding that helps to make it easy for mothers to want to look after their

The baby orang-utan above was born much more able to fend for itself than a human baby – its grasp is strong enough to support its own body weight while hanging on to its mother's hair. Similarly a newborn springbok (below) can climb to its feet and walk unaided within minutes of birth.

Even if, by some miracle, a human baby could stand, walk and hang by its hands at birth, he or she would still be a baby culturally speaking. The purpose of our prolonged juvenile state is the acquisition of culture, all the mental and physical skills which enable us to become viable and reproductive members of human society.

children, and these immature characteristics also persist much longer in humans. Indeed, these features may be responsible for another type of caring in our society – attractiveness of women to men; childlike complexion, roundness of body form and unbroken voice are all characteristics that are seen as more feminine than masculine.

One other characteristic that reinforces parental care is weeping, unique to the human infant. For a baby who is away from his mother's arms, this is the only way of attracting attention, and it usually produces very quick results.

In much the same way that we traced the development of the eye, we can trace the progress of human development in our animal ancestors. The increasing tendency to be born at an earlier stage leads to a greater period during which we are out in the world but protected from the need to fend for ourselves. During this protected and cushioned time of life, young humans learn voraciously. One important way they learn is by playing. Play is a way of interacting with the world to find out how it works, a way of rehearsing for adult life. And it provides fertile opportunities for using the imagination. All of these are skills which can lead to innovation, initiative and creativity if they persist into adult life. If, therefore, we wanted to look towards the future evolution of human beings perhaps this process of drawn-out development provides a clue.

It may well be that we will live longer in the future. But if we do it is likely to be as a result of a stretching out of our whole pattern of development, rather than continuing to age as at present but living to 120. In the words of J.B.S. Haldane, a major contributor to the study of human development; 'If human evolution continued in the same direction as the immediate past, the superman of the future would develop more slowly than we, and be teachable for longer. He would retain in maturity some of the characteristics which most of us lose in childhood... He would probably be more intelligent than we, but distinctly less staid and solemn.'

Perhaps, then, the human being of a few millennia hence would not look particularly strange to our eyes. To be transported to a town of the future might show us a community in which learning and play had a much greater part, and where major technological and intellectual achievements had taken place. But the people would look the same. Until, that is, we realized that the toddler waddling across the grass was 12 years old, the sturdy girl playing with her computer was 35, and the teenagers had been born 50 years before.

Of course this can only be intriguing speculation, but it has a very plausible ring to it. One important characteristic of the future human adult will probably be a greater involvement in play for play's sake. And who can deny the importance even nowadays, among people who are not ashamed to confess it, of a playful attitude to life and the world? From sports to computer games, humour to crosswords, we humans are playful animals and may well become more so in the future. That is one possible future for us, and a rather more pleasant one to close this book with than other potential futures that are currently put before us.

General

Carlson, Neil R. 1981: *Physiology of Behaviour* (2nd edition). Allyn and Bacon Inc.

Despopoulos, Agamemnon and Silbernagl, Stefan. 1981: *Color Atlas of Physiology.* Year Book Medical Publisher Inc.

Durkin, Ned. 1979: *An Introduction to Medical Science.* MTP Press.

Ganong, William F. 1981: *Review of Medical Physiology* (10th Edition). Lange Medical Publications.

Lamb, J.F., Ingram, C.G., Johnston, I.A. and Pitman, R.M. 1980: *Essentials of Physiology.* Blackwell Scientific Publications.

McClintic, Robert J. 1980: *Basic Anatomy and Physiology of the Human Body* (2nd edition). John Wiley and Sons.

McNaught, Ann and Callander, Robin. 1975: *Illustrated Physiology* (3rd edition). Churchill Livingstone.

Romer, Alfred Sherwood and Parsons, Thomas S. 1977: *The Vertebrate Body* (5th edition). W.B. Saunders Company.

Young, J.Z. 1974: *An Introduction to the Study of Man.* Oxford University Press.

Specific topics

Austin, C.R. and Short, R.V. (ed.) 1976: *The Evolution of Reproduction.* Cambridge University Press.

Carpenter, R.H.S. 1984: *Neurophysiology.* Edward. Arnold.

Coren, Stanley, Porac, Clare and Ward, Lawrence M. 1979: *Sensation and Perception.* Academic Press.

Geschwind, Norman. 1964: *The Development of the Brain and the Evolution of Language* from Stuart, C.I.J.M. (ed.) Monograph Series on the Development of Language, No. 17.

Gowitzke, Barbara and Milner, Morris. 1980: *Understanding the Scientific Bases of Human Movement* (2nd edition). Williams and Wilkins.

Hamilton, David and Naftolin, Frederick (ed.) 1982: *Basic Reproductive Medicine.* The MIT Press.

Montagu, Ashley. 1981: *Growing Young.* McGraw-Hill.

Oakley, David A. and Plotkin, H.C. 1979: *Brain, Behaviour and Evolution.* Methuen.

O'Riordan, J.L.H., Malan, P.G. and Gould, R.P. (ed.) 1982: *Essentials of Endocrinology.* Blackwell Scientific Publications.

Popper, Karl R. and Eccles, John C. 1977: *The Self and Its Brain.* Springer International.

Sadow, J.I.D. 1980: *Human Reproduction.* Croom Helm.

Sanford, Paul A. 1982: *Digestive System Physiology.* Edward Arnold.

Tanner, J.M. (ed.) *Control of Growth.* British Medical Journal 9999.

Viidik, Andrus (ed.) 1982: *Lectures on Gerontology: Volume 1: On the biology of ageing.* Academic Press.

Index

Artwork

Cynthia Clarke
18, 21, 22, 22-3, 32, 34, 35, 38, 40, 47 top, 57, 58-9, 88, 92, 94, 96, 97, 101, 103, 108-9, 119 right, 121, 130 bottom, 133, 135, 138 bottom, 142, 174, 177, 182, 196, 197, back cover
Cooper West 37, 42, 45, 46, 48 left, 60, 63, 69
Cucumber Studios 18-19, 38 top, 41, 47 bottom, 48 right, 50, 53, 74, 74-5, 83, 91, 105, 119 left, 138 top, 141, 162, 168, 174 top, 184
Tom McArthur 44, 56, 70, 126, 150 left, 170
Janos Marffy 28 left, 31, 33, 75 right, 85 left, 100, 127, 128, 130 top, 136 centre
Mulkern Rutherford 16, 19, 27, 54, 73, 75 left, 76-7, 85 right, 98, 144, 145, 150 right, 154, 180, 181, 186, 190, 199, 200 bottom
Charlotte Styles 28 right, 200 top, 202, 209
Tassos Xeni 211

Photographs

Mike Abrahams/NETWORK 61, 205 left
All Sport 94-5
Clive Barda, London 87 top, 107
BBC Hulton Picture Library 62 top right, bottom right
Biophoto Associates 10 left, 25 right, 48 bottom, 63 bottom right, 67 bottom, 69 top right, 71 top right, bottom, 73 bottom right, 81 inset, 92 bottom centre, 105 top centre, 128 bottom, 129 top left, 149 centre, 151 left, 152 top left, centre left, centre right, 153 bottom, 157 bottom left, bottom centre, bottom right, 159, 162 bottom, 163 left, 168, 169 bottom, 177 bottom, 180, 182 bottom, 190 right, 208 bottom left, 210 left
Catherine Blackie 131 pictures 1-5
Blackwell Videotec 21 top left, 23, 36, 44, 58, 63 left, top right, 65, 66, 73 top left, 96, 97, 98, 122 top left, 138 top, 152 bottom, 155 bottom left, 194 top right, 204 top left, 208 bottom right, 211 top left
Austin J. Brown/Aviation Picture Library 79 top, 84
Camerapix Hutchison 148
Camera Press 62 top left
Charing Cross Hospital/A. R. Williams 196-7, 211 top right
CNRI 73 centre right, 179 right
Bruce Coleman Limited 10 top right, bottom, 11, 12-13, 13, 100, 134 left, 217 top left, top centre, bottom
Colorific! 90 right, 129 top right, 142 bottom left, 146 right, 149 left, 187 left, 189 bottom left
Cranston-Csuri Productions Inc. 136 top
Daily Telegraph Colour Library 38 top, 192-3 bottom
Theo Davies/University of Exeter 115, 118 right, 121 top, 123 bottom
Zoe Dominic 73 top right
John Frost Historical Newspaper Service 85
GLC Fire Brigade 131 picture 6
Goldcrest front cover, 30 bottom right, 55 top, 59, 64-5, 67 top left, top right, 92 top left, 106, 112, 121, 146 centre, 148 inset, 155 top left, 169 top, 173, 177 top, 178 top, David Barlow 28 centre, 42 bottom left, 46, 53, 55 bottom right, 57 bottom, 80, 88, 92 bottom left, 105 top left, 108 top, 114, 135 top, 138 bottom, 142 top left, 145 top left, top right, 147 centre, 149 right, 158-9, 160-1, 161 top left, bottom left, bottom right, Herbie Knott 20 centre, 87 bottom left, 89 right, 92 top right, 102 left, 178 bottom, Dr Frances McKirton, Middlesex Hospital 39 left, right, Roger Pederson 190 left

Gower Medical Publishing Ltd. 30 top left, centre, 41 centre, 43, 57 top, 81, 92 bottom right, 105 bottom left, 122 bottom left, bottom right, 142 bottom right, 145 bottom, 171 left, 183 right
Sally & Richard Greenhill 27, 207 left, 218
Hammersmith Hospital 32 bottom left
Camilla Jessel 204 bottom left, bottom right
Frank Lane Agency 217 top right
Mansell Collection 71 top left, 89 left
Marshall Cavendish Ltd /Ron Sutherland 205 right
Leo Mason 64 top left, 124-5, 140
Multimedia Publications 213 right, Balnicke 69 top left, 110, 146, Herbie Knott 44-5
Petit Format 32 top, centre, 73 centre left, 77, 82, 109 top, 117 left, 174, 182 top, 183 left, 185 centre right, 189 top left, right, 191, 192 top left, top right, 192-3, 193 top left, top right, 195, 199, 202 top, 202-3, 204 top right
Picturepoint 134 right
Playboy Television Limited 185 left
Popperfoto 62 bottom centre
Rex Features 64 top right, 213 left
Ricordi 106-7
Francis Ring 105 bottom centre, 118 left, 123
Science Photo Library 21 bottom right, 24, 30 bottom left, 32 bottom right, 32-3, 38 right, 42 bottom right, 48-9, 50, 64 bottom right, 69 bottom right, 73 bottom left, 105 top right, 125, 147 top, 164 top, bottom right, 165 top right, bottom, 166, 171 centre, right, 185 top, 194 bottom right, 215 left, Dr. Tony Brain 135 bottom
Frank Spooner Pictures/Gamma 214
Tony Stone Worldwide 52-3, 90 left, 90-1, 132-3, 139, 181, 187 right, 194 left, 208 top left
Syndication International 163 right
Topham 62 bottom left, 111
University College Hospital/Dr Joan Round 101
Vautier de Nanxe 29 left, 87 bottom right
Vision International 102 right, 128 top, 151 right
John Watney 17, 35 top, 38 left, 42 top, 51 left, 71 centre, 79 centre, 123 centre, 127, 147 bottom, 151 centre, top right, 157 top, 158, 161 top right, 193 bottom, 210 right, 211 bottom, 215 right, 216
ZEFA 14-15, 24-5, 29 right, 35 bottom, 51 right, 55 bottom left, 116, 117 right, 121 bottom, 122 top right, 155 top right, 202 bottom, 206-7, 208 top right

The pictures on the endpapers, pages 18, 21 bottom left, 58-9, 153 top, 168-9, 179 left and 184 are from TISSUES AND ORGANS: A TEXT-ATLAS OF SCANNING ELECTRON MICROSCOPY by Richard G. Kessel and Randy H. Kardon. W. H. Freeman and Company. Copyright © 1979.

The pictures on page 113 are from DREAMSTAGE Scientific Catalog © J. Allan Hobson and Hoffman-La Roche Inc.

Multimedia Publications (UK) Ltd have endeavoured to observe the legal requirements with regard to the rights of the suppliers of photographic and illustrative materials.